The Complete Guide To
DIRT STOCK CAR FABRICATION AND PREPARATION

By Joe Garrison, GRT Race Cars

ISBN # 0-936834-95-1

Editor: Steve Smith
Editorial Assistant: David Wolf

Published By

P. O. Box 11631 / Santa Ana, CA 92711 / (714) 639-7681
www.ssapubl.com
Manufactured and printed in USA

Table of Contents

Chapter ... Page

Chapter ... Page

Thanks

This book could not have been written without the knowledge shared with me throughout the last 14 years. This knowledge was provided by some of the best people involved in racing. I would like to extend a very special thanks to: Scott Keyser and all of the fine people at AFCO, Russell Baker, Jeff Scales, Kenneth Taylor, Janis Hopper with Kilsby Roberts, Sam Garrett with CARS, Lee Sales with TCI Transmissions, Bert Transmissions, John Brunson, Bill Behling, Johnny Stokes, Tom Guithues, Terry Phillips, Larry McDaniels, Willie Sellars, Jeff Jackson, Tom Barkley, Tim Macomber, Jack Sullivan, Alan Rippy, Bill Hooten, L.T. Davis, Shane Lauderdale, Shane Burrows, Gary Burrows, Tony Cardin, and all the GRT Race Cars employees. I sincerely appreciate all the time you have taken to help me.

Also thank you to Bill Frye and his wife Carol, who have helped build GRT to what it is today, Johnny Virden, Freddy Smith, Gary Webb, and Jerry Inmon, who have worked with me very closely for many years.

I would like to thank all of the GRT Race Car drivers and crews who have helped build the "winning tradition".

Special thanks go to Jamie Lauderdale, my secretary, who typed every word of this book, and to Steve Smith, who made the commitment to publish this book, and who spent hundreds of hours editing the manuscript.

Most importantly, a very special thanks goes to my wife, Sheila, and daughter Randa Jo, and my step son and daughter, Jeremy and Jayme Roberts. They have stood by me throughout my racing career. Also, a special thanks to my brother, Mike, who was originally a part of GRT, to my mother, who has supported me always, and the Good Lord who allows all things to be possible.

In 1995, two very special people that were close to me passed away: my father-in-law, Charles T. Sublett, who inspired me and gave me the opportunity to build my race car business, and my dad, Joe Garrison, Sr., who was the real driving force behind me because he taught me to work hard and to never give up. In memory of them, I would like to dedicate this book. Thanks for all of the faith and support you gave.

Joe Garrison, Jr.

Disclaimer Notice

Introduction

Engines, fast cars, and racing was all I knew. Building cars and trucks was something that naturally came to me. Then one Saturday night in the summer of 1981, I attended a dirt track stock car race. I really can't describe what happened at that moment except for the fact I knew stock car racing was where I was going to be.

Each week I attended the local track and some various distant tracks. Soon I approached a driver in my home town and we agreed on a partnership for the upcoming season. I saved my money all Winter and came up with my half ($2500.00) to buy a complete race car (motor and all). We used an old flat bed trailer, built a tire rack and worked hard all Winter preparing for the next year. Through many ups and downs that year I learned that all the money I could make, all the time I could spare and all the help we could find was just the beginning of what it takes to race a stock car successfully.

I made my decision and I sacrificed about all I had including my job to pursue a career in stock car racing. From then on I dedicated myself to dirt track racing's highest level, late model dirt cars. It was just a dream at this time.

My first big break was an opportunity to participate on one of the sport's best teams. Larry Phillips gave me the opportunity to go to Daytona and participate in Speed Weeks. This was it, the ultimate, the best there is in dirt late models. We were racing against Purvis, Moran, Swartz, Boggs, Smith, Combs, Moore, Bloomquist, and Moyer.

When we arrived back home, I approached a local chassis builder with a proposal to form a team and race a house car operation for him. He said, "you pay retail just like anybody else." With two seasons already gone by and 35 feature wins in limited late model in his cars, I thought he would be in favor of this but he was not. We could not afford to buy a late model so there was only one way to go — build our own!

This proved to be a major decision because of the disadvantages we already had. Where do we start? What do we do? Now we had to build the cars, maintain the car and race the car on a limited budget. We had to do something to offset this and the only thing left was offer the race cars we were racing for sale. So now what do we call them? Our Garrison Racing Team was now selling chassis, so we shortened the name and became GRT Race Cars.

Many things have happened from that day. We've been involved with many drivers and teams to promote our cars. Our first real serious effort was in 1987 with Wayne Brooks driving our house car in which we won fifteen features in thirteen states and finished second in NCRA points and the New Memphis Motorsports Park late model division. This was done with one car and one motor. Man we were really having fun!

With the 1987 season ending we had just made the first impression on many drivers and teams throughout the country. What we were about to find out through years to follow was that gaining respect from top racers was a major job, especially with so many established chassis builders throughout the country already. We had a fairly successful year and established our slogan which we still use today: "We Build Winners."

Through the following years we worked with several different drivers on a house car basis — T. J. Paushert, Jerry Inmon, Eddie Pace, Johnny Stokes. We even had the luxury of having an employee, Tony Cardin, who helped build the cars at the shop drive one of our house cars. These drivers and many others had a lot to do with the success of promoting and developing our cars.

1987 was the first design car that was built and it was drawn on a 4 x 8 sheet of metal. Then through the early years with all of the drivers and the constant on-track testing, and feedback from our customers, we kept developing and improving our original car design.

Along with this we were also working with one of the finest parts and fabrication shops in the industry, TWM from Louisiana. They would take our blueprints or designs and build all the components for our cars. We also worked with TWM on the development of the aluminum bird cage still being used today.

So now we were racing and building a business. The responsibility of building a race car, working with the racers and owners, and managing a full time race team was becoming a burden. I could not concentrate on the two businesses. I needed an in-house team that could manage, develop, and promote GRT Race Cars so I could take care of the race car business. I needed to be at races working with the racers, answering the phone and working with other manufacturers to promote and learn their products. The next turn of events gave me the opportunity I was looking for.

I needed a self-managed, dedicated, hard working in-house team that could work hand-in-hand with me on design and development and promote the company while I tended to taking care of the business end. I made an offer to Bill Frye, who moved from Missouri to Arkansas and became part of the GRT Race Cars team.

The year was now 1991 and the next four years developed into what GRT is today. Frye joined me in a goal to be one of the best racers in the nation.

Today as you read this Frye is one of the top contenders in dirt late model racing and GRT is the upcoming race car that can be backed up with years of accomplishments and some of the best teams and drivers in the business.

Along with our growth and success we also involved ourselves with an IMSA road racing team (Charles Morgan and Sam Garrett doing business as CARS) that ran their program out of one of our shops. They have ventured into the computer technology end of design and development of race cars, and with their help we are currently designing new front end geometries and chassis designs. Combine all of this and put years of experience together and what you have is a company that dedicates itself to dirt track racing from IMCA to late models.

In closing this introduction I would like to leave with you some of the accomplishments and race teams that enabled GRT to write this book:

Five Time MLRA Manufacturer's Champion

Four Time S.U.P.R. National Champion
Doug Ingalls
Tony Cardin
Donald Watson
Ronnie Poche'

Three Time MLRA National Champion
Bill Frye
Terry Phillips
Alan Vaughn

1994,1995 Deery Bros. Champion
Gary Webb (53 Feature Wins)

1994 Four Crown Champion, Eldora Speedway
Freddy Smith

1994 Hillbilly 100 Champion, Pensoboro Speedway
Bill Frye

1995 6th Annual Hav-a-Tampa Shootout Champion
Bill Frye

1995 Super-MLRA Shootout Champion
Ronnie Poche'

1994 Topless 100 Champion, Batesville Speedway
Tony Cardin

1995 UMP Champion
Tony Izzo Jr. (31 Feature Wins)

1995 S.U.P.R. Champion
Ronnie Poche'

1995 MLRA Champion
Alan Vaughn

1995 Southern All Star Champion
Marshal Green

1995 Hav-a-Tampa Manufacturer's Champion

1995 Hav-a-Tampa Rookie Of The Year
David Gibson

1995 Sunoco Late Model Series Champion

Kris Patterson

1995 Grand Casino Late Model Series Champion
Tony Cardin

1995 NCRA Late Model Champion
Larry McDaniels

1995 UMP Chassis Manufacturer Champion
100 wins

Joe Garrison
GRT Race Cars

Chapter

1

Design Of A Race Car

When designing a race car, you have to consider the type of racing you are going to be doing, the speeds you will be running, and the most common track sizes. You must also consider the weight rules for the type of racing you will be doing, never letting safety out of the picture. Other considerations are ease of maintenance, structural strength, basic suspension design, and safety.

Design For Ease of Maintenance

A major consideration of race car design is making the car easier to work on during maintenance or general repair. For instance, all of our cars come standard with bolt-on front bumper and passenger side door bar, and rear bolt-on bumpers. What this does is make repairs at a race track or in the shop much simpler and faster. Whenever you don't have to get the torch, welder, and Porta-Power out, things can be done much more efficiently. Also, we make our body bracing and supports out of 1/4" x 1" aluminum strap so that when impacts are taken, the aluminum strap absorbs the blow. The strap can be straightened without using the welders and torches. Also, the bolt-on application of any component that possibly might be in a crash zone simplifies repairs.

Other ways of making working on a car easier are using Dzus fasteners on as many panels as possible. This includes removable fenders for easy access to the engine bay, removable dash panel for easy gauge and brake pedal maintenance, and removable dry sump tank panel for ease of changing oil or checking the in-line oil filter. Some cars even have the whole body side removable for easy changes or repairs.

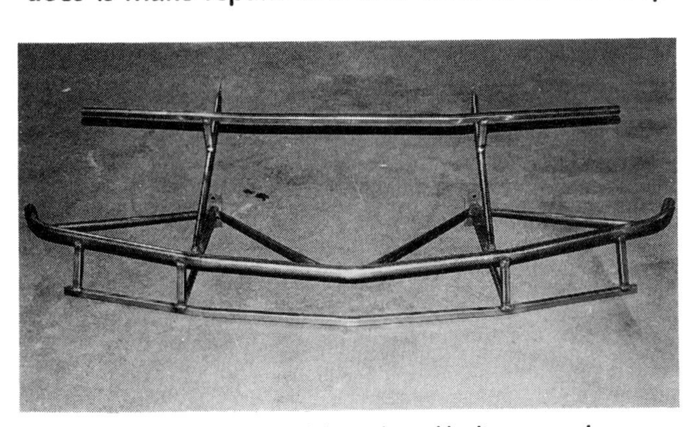

A bolt-on front bumper (above) and bolt-on rear bumper (right) makes repairs much simpler and faster. They help to form a crush zone that minimizes damage to critical parts of the chassis, as well.

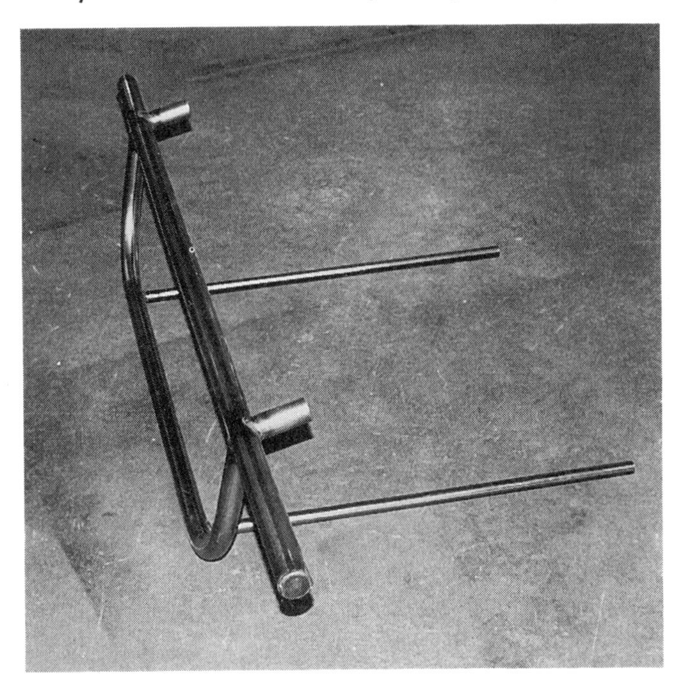

The cage and chassis should be designed for ease of maintenance for all components.

The fuel cell placement can be moved up or down or left and right to take advantage of the weight mass.

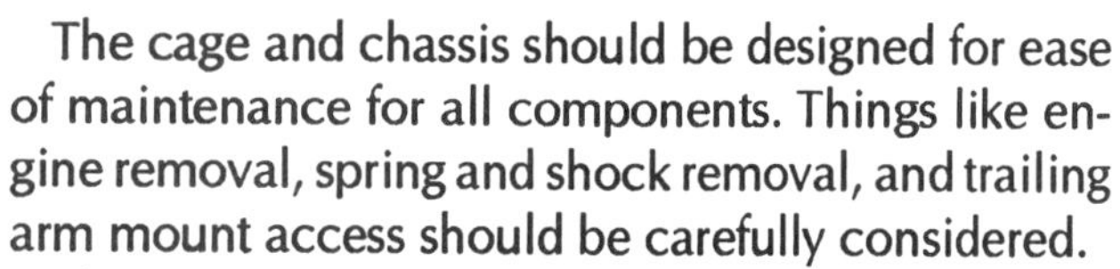

The cage and chassis should be designed for ease of maintenance for all components. Things like engine removal, spring and shock removal, and trailing arm mount access should be carefully considered.

There are several design considerations that go into a race car that GRT builds for easy maintenance and repair simply because these factors also improve a race team's chances of getting back on the track after repairs or changes as soon as possible. You don't need to miss a race or a hot lap just because of a minor problem that could have been easy to fix with just a little bit of design consideration for making a car easier to work on.

Choosing Component Location

When designing the chassis many things have to be considered. For instance you must incorporate the bracketry and mounting locations of all your suspension and bolt-on components around the structural design of the car. Sometimes putting your brackets and mounts where they belong is a problem because of the location of the roll cage tubing or bracing that is located there for optimal strength. In this case you possibly might increase the wall thickness or outside diameter of the tubing and change the location to accommodate the bracket and still retain the same chassis stiffness.

One general rule for component placement is that you want to use each and every piece you install in a chassis to the best advantage possible. For instance, most items placed in the chassis are put there to take advantage of the weight. When putting weight in a chassis you try to make use of it as much as possible. These component choices include the battery, dry sump tank, and fuel cell (it has to be at the rear but can be moved up and down or left and right).

Well-planned component placement can also help one piece to serve dual purposes. For instance, placement of a motor plate or torque plate serves two purposes — to hold the engine in location and to strengthen the chassis.

Component placement in a chassis has to be fairly practical and is determined by taking advantage of weight and the size and function of the component.

The torque plate can serve two purposes – to locate the engine and to strength the chassis. It serves as a triangulated structural member of the chasis.

For optimal chassis strength, triangulation must be used in the design. Notice the abundant use of triangles in this chassis.

Designing For Structural Strength

When bracing a chassis for optimal strength, triangulation must be used, and the more triangulation you have, the stronger your chassis is. You want your chassis to be strong and light, but never sacrifice weight for safety and durability. You can't win races if your car is not durable, and a lightweight car can get you hurt.

The basic principle of triangulation is to spread the loads which are input to the chassis evenly throughout the entire structure. This is done to place the loads in the tubes in compression rather than in bending, which utilizes the strongest aspect of round tubing.

A major design consideration for structural strength is the wall thickness and the outside diameter of the tubing used. Increasing the O.D. of the tubing makes the point of failure much higher than just increasing the wall thickness. For instance, if you do a mathematical analysis of a 4-foot long piece of 1.75" O.D., .095" wall tube versus a 6" O.D., .065" wall tube of the same length, you will find that the larger diameter tube with a thinner wall thickness is actually two times stronger in both critical bending stress and deflection strength. And, the larger diameter tube weighs only 2.45 times more. This is an extreme example, but it demonstrates how an increase in diameter increases a tube's strength in all directions (compression, torsional and bending).

As a practical example, a 4-foot length of 1.75" O.D., .095" wall tube and the same length of 2" O.D., .065" wall tube has the same strength, but the length of 2" O.D. tubing weighs 1.4 pounds less !

Designing For Weight

Another design consideration is that the rules in most sanctioning bodies require you to weigh enough that you usually have to load your chassis down with ballast, so you have plenty of room to increase the strength and safety of your car without going over the weight limits.

As an example of this, late model dirt cars weigh about 1,950 pounds on average and the weight of the driver averages about 200 pounds, so that totals 2,150 pounds. Most rules require a car to weigh in the 2,300-pound range, so you have plenty of room to put ballast in the proper place. As long as you have this advantage, then it makes sense to add a few additional tubes in the chassis design to help make sure the area between the front and rear suspension attachment points is absolutely rigid.

Establishing Roll Centers

We try to establish our front roll center laterally as close to the center line of the car as possible, and about 3 inches to 5 inches above ground level. This is a critical area. If a roll center is too high or moves around a lot during body roll, the car is not consistent and handles poorly. What we like to see in a front roll center is the roll center staying as consistent as possible when the car is under braking and rolling to the right as the car goes into the corner. If it does move around during corner entry, it should move to the right gradually as the car rolls. This will stiffen the right front to some degree and help the car get back on the left when leaving the corner.

A roll center that is too high will shear the front tire patch and cause a push condition. In our opinion, 5 inches is the maximum front roll center height. Some cars are built with a suspension that can adjust the front roll center height in order to adjust the car to track conditions. Being able to have a front roll center for certain track conditions is an advantage, but is not recommended for the amateur racer. We build our cars with a somewhat universal roll center that works well on most track conditions.

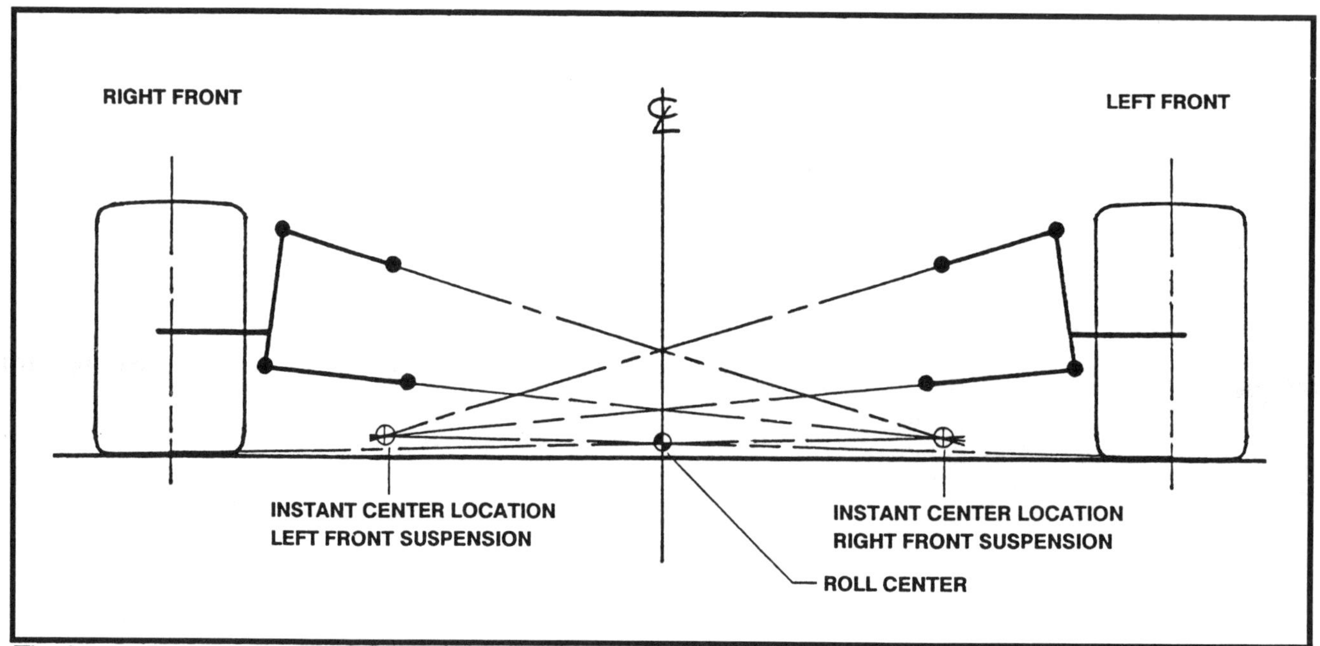

The front roll center can be found by extending lines through the control arms on both sides to a common intersection point, called the instant center. Then a line is projected from the instant center back to the center of the tire. The roll center for that side of the car is that point where it crosses the center line of the car. The true front roll center of the vehicle is where the instant center swing arms of the left front and the right front cross each other.

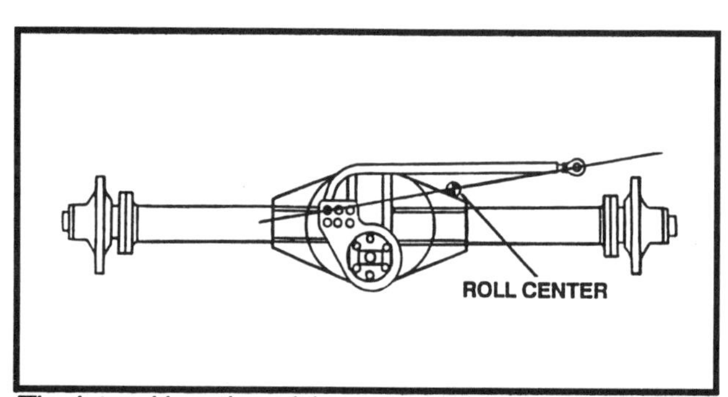

The lateral location of the rear roll center is the absolute center of the Panhard bar. The roll center height is that distance above ground.

Rear roll centers can be anywhere from 10 inches to 16 inches above ground, and somewhere in between works well on most applications. Rear roll centers are established by the Panhard bar location on the car. All of our cars have the Panhard bar mounted on the left hand side of the chassis and they run downhill to the rear end mount on the right side. Running the Panhard bar downhill at an angle creates what we call mechanical side bite. This induces roll into the car not only with the roll center location but with the angle of the bar (see more details on this in the Rear Suspension chapter).

The length of the Panhard bar determines the lateral roll center and the severity of the roll produced. The lateral roll center is located halfway between the two mounting points of the bar. The height is measured from the ground to the center of the bar.

The front roll center location is a lot more critical than the rear roll center because the front of the car handles 70 to 80 percent of the total rolling force of the car. Once the front roll center is properly established, the rear roll center height can be used as a fine tuning adjustment for the chassis.

Front Spindle Design

Front spindles need to be designed around the needs of your chassis. In the past we have used three different types of spindles with three different types of front ends. The current design spindle has an advantage in helping determine front roll center location because of the overall length of the spindle. In other words, the length of the spindle combined with the suspension mounting points determine the roll center.

Another design factor with the spindle is the steering arm placement. This helps determine several

The spindle design helps determine front roll center location, bump steer, steering ratio and Ackerman steer.

things like the bump steer adjustability, the quickness of the steering ratio (determined by the length of the steering arm) and the amount of Ackerman steer. (See the Front Suspension chapter for an explanation of Ackerman Steering.) With the Ackerman, we like to see as little as possible on our cars. We have about one degree to two degrees of toe out when the wheels are turned twenty degrees.

Establishing Camber Gain Curves

We like to see approximately one degree of negative camber gain per inch of shock travel on the right front, and we like to keep the left front as close to zero camber gain as possible. The left front camber gain is not possible to do and also have a roll center that you can live with, but you can get close. Trial and error is part of the design process.

Camber gain can be changed with A-arm length and location, but keep in mind that changing lengths and location affects roll centers again. So it's a time consuming process and is designed into our chassis for the best possible performance.

Watching tire wear on your tires from inside to the outside of the tread width can tell you if you are close on static camber setting. If you can't get your tire wear even on your fronts by changing static camber, then most likely your camber gain is way off.

Designing The Bump Steer

We like to see the bump steer on one of our cars at .015" to .020" toe out per inch of shock travel on right front. The left front bump should be as close to zero as possible. These specifications are our opinion and we feel maximum performance can be obtained with these specifications.

The bump steer adjustments are made by shimming the tie rod up or down at the spindle and the rack end of the tie rod. The spindle steering arm design and the rack mounting plate placement gets the bump steer close. If these two mounts are not close, bump steer will be extremely hard to set. These are the two main provisions we put into the steering design to make bump steer easy to set.

Computer-Aided Chassis Design

GRT took its current chassis design and ran a computerized chassis analysis on the structural design.

We used a chassis structure design program called FEA (Finite Element Analysis) by Algor. This program works by entering the actual shape of the chassis into the computer as a stick drawing. You enter in the front, rear, top, bottom, left and right sides of your chassis. At the same time you must enter the OD size, the wall thickness, and overall length of each and every piece of tubing in the chassis. Then the chassis can be tied down in the computer drawing at three suspension input points on the chassis.

Let's say both rears and the left front are tied down. Then a specified amount of force is applied to the chassis pulling down and pushing up. This produces a certain amount of twist and flex in the chassis. The computer program then tells you how much force it took to flex the chassis a certain amount. For example, say 4000 ft. lbs. of force produced one degree of twist. The program will then show the areas of the chassis that are under high stress, medium stress, and lowest stress. This enables you to start adding or taking away tubing, determining which tubes need to be there and which ones do not.

This process takes a lot of time with trial and error involved. The program takes into consideration not only the location of all tubes, but also the diameter and wall thickness of tubing. You have to work with these two elements to determine the maximum strength versus minimum weight.

Using an analysis type of program with a computer can help to rid the chassis of virtually all twisting and flexing.

After you get the right amount of tubing in an area that needed attention and determine the right size, the next time you run the analysis it will show the new readings and the improvements in all of the high, medium, and low stress areas. The program tells us what the chassis is doing that the naked eye would never detect.

We decided that we wanted to increase the rigidity of the chassis, so we went to work on the computer. After many hours of trial and error we found that changing the location of several intersecting points, and tubing O.D. and wall thickness, the structural strength of the new car was 3.7 times more rigid and we only added 32 more pounds of total weight. What a difference for so little weight gain! Thanks to the computer we found this out. We probably would have never known what was required to gain this strength without the use of computer analysis.

Some people argue the fact that a dirt car needs to have a certain amount of flex in the chassis design, but our current cars are structurally rigid and work well. We decided to improve the design and have a more tuneable and predictable chassis which comes with having a stiffer chassis.

Building In Crash Zones

We now have a strong and light chassis, so we had to build in some forgiveness in the front and rear sections of the chassis to help prevent the car from being totally destroyed in case of a crash. This also gives the driver an additional safety factor of having crush areas to absorb heavy impacts. Naturally the front and rear bumpers and right side door bars are braced good enough to hold the body components and take normal racing abuse, but they are light enough to crush on impact to absorb the crash. This in turn helps keep the main chassis from being hurt and helps protect the driver.

Crash zones are created by building a structure like a front bumper assembly that is very strong as far as holding the nose and body panels in place, but it will deform under heavy impact. It has a certain amount of triangulation using a smaller diameter and thinner wall tubing. For example our bumpers are made out

This is the GRT chassis after being redesigned by the computer analysis program. Notice the extensive amount of triangulation used in the chassis design.

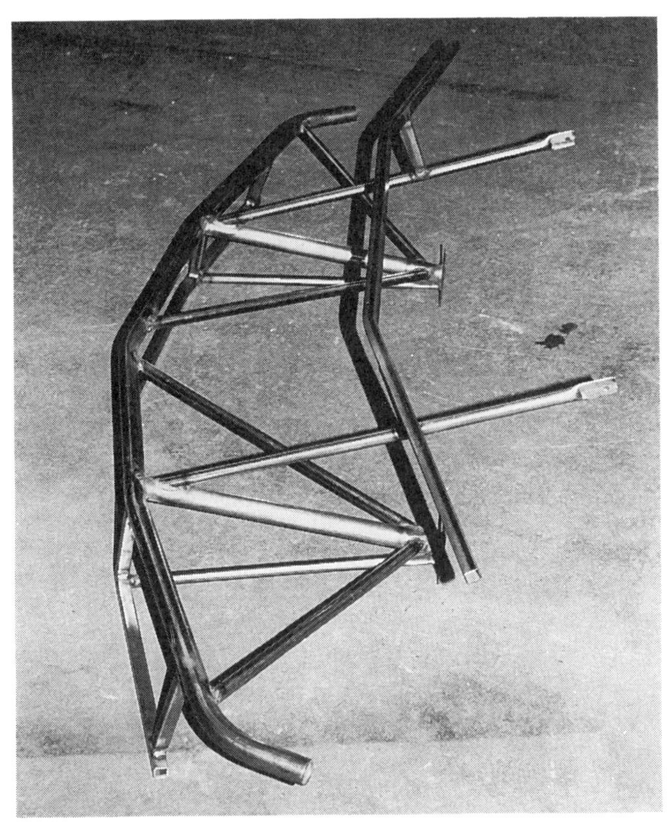

Crash zones are created by using a structure like this front bumper assembly which is strong enough to hold the body panels in place, yet will deform under heavy impact.

of 1.25" O.D. x .065" and .75" O.D. x .065" tubing. These are relativity light gauge tubing and hold up well during normal use. In the event of a crash, however, this bumper will collapse and absorb much of the frontal impact. This saves the main structure and gives the driver a cushion of safety.

In addition to the bumper, the actual front area of the bay from around the radiator to the front shock mounts should be considered a crash zone. There is no need for very stiff structural rigidity there. The chassis should have no triangulation there so it will collapse with relative ease in case of a severe frontal impact, providing even more of a crash zone.

The rear clip or tail section is designed the same way. It has 1.25" O.D. x .065" bracing and is built to absorb rear impact and not hurt the main structure of the chassis or affect suspension points. Some of the ways this is done is by using some of the bracing in the chassis just where it is needed. What I mean is the down bars from the top hoop are not run all the way back to the rear bumper like many chassis designs I have seen. This keeps the entire chassis

(Above and below) A crushable box type of structure is used behind the attachment points of the down bars to absorb rear impact and not hurt the main frame structure.

from being affected by a rear impact because these two down bars are tieing everything together. When you move these attachment points forward and use an under slung frame to build the tail section, it prevents the chassis from taking abuse forward from the attachment point of the down bars. Also, from this point to the rear you have a strong but very light and crushable box type structure that holds your body mounts, fuel cell, and rear bumper in place (see photo). In general what you have is a box that is strong and yet collapsible.

This same type of engineering is built into the right side or passenger door bar. It is built with 1.25" x .065" and .75" x .065" tubing so that in the event of a side impact it will collapse and cushion the blow for the driver, and have minimal damage to the main

Sturdy, triangular type of driver's side door bars help to add side protection to the driver.

frame. This bar is as strong as it needs to be for holding the body and still is collapsible.

The driver's side door bars are another story. Safety is a key factor and you can't sacrifice strength here. Driver door bars should be designed so they hold up under severe impact and protect the driver. In our opinion triangular type door bars produce maximum strength (see photo).

Crash zones should be engineered into the chassis to try to keep the structure of the chassis as raceable and repairable as possible in case of impact. Driver safety is the most important factor, and must be the top consideration.

Remember that these crash zones must be determined by how much the car weighs and the speeds that are being run. Heavier cars and higher speeds require that the crash zones be engineered differently with different sizes of materials.

Cockpit Safety Design

The driver's compartment must be designed so the driver won't injure himself in case of an impact or a roll over. You need plenty of room around the driver with no blunt objects in reach of him, plenty of head room and door bars that are triangulated away from the driver for maximum side impact resistance. Think of all the ways that another car could possible impact your race car, such as driver's side, another car landing on top of your car, etc. Then design the cockpit layout to provide plenty of safety for the driver in any type of condition you could imagine.

Design Is The Key Element To Being Competitive

The design of the car should be the strongest, lightest, and most functional that it can be. The design must have all of these features to be a winner. When you build in strength, you build in reliability. When you build in lightness, you build in quickness. When you build a functional race car, you have a simple, adjustable, and practical race car. These are key elements to a winning race car chassis.

Chapter
2

Race Car Construction

Construction of a race car chassis starts in the tubing rack. From there you must have a good plan to pre-cut and bend all of your tubing before you start constructing the chassis in the jig.

A good sturdy jig with removable attachment points is the best way to assure a quality chassis. You must carefully level the jig side to side and front to back. Pieces must be securely mounted to the jig fixture to hold firmly while you start to tack-weld the frame together. We usually tack-weld our chassis together in two to three locations small enough that when you run your main weld bead over the tubing, the tack weld is not visible. Starting and stopping the weld also should be done in a manner where the outer lap looks like one continuous weld.

A good sturdy jig with removable attachment points is the best way to assure a quality chassis.

Chassis Building Materials

Before we continue with the way to properly construct a chassis in a jig, let's talk about the types of materials we use. We have found that .083-inch wall Drawn Over Mandrel (D.O.M.) mild steel round tubing is the best for our application. Our 2 x 2 square tubing uses an .083-inch wall thickness also. Some customers prefer us to build cars out of a thicker wall tubing, and that is not a problem.

The grade of steel that we most commonly use is 1020. It is not the strongest of the low carbon steels, but it offers all of the specifications required for a race car building application. The weaker grades of tubing bend real easy and don't have good return properties. Each number such as 1010, 1018, 1020, 1026, 1040, etc. designates how much carbon content is in the tubing. The higher the last two numbers of the four digits, the higher the carbon content of the steel. Naturally the more carbon content, the stronger the steel.

D.O.M. steel tubing is formed from a steel strip. It is electric resistance welded, and then the welded tube is cold drawn over mandrels to create a smaller diameter and thinner wall tube. The cold drawing process works the seam weld, making it virtually disappear into the metal structure of the tube. It is this cold drawing process that gives D.O.M. tubing a superior tensile strength and yield point over tubing manufactured from other processes. For instance, D.O.M. tubing has double the yield strength of hot finished seamless tube.

Also an option on building a chassis is using 4130 chrome-moly tubing. However, we find that cars built with 4130 tubing are not as forgiving in crash

MIG welding uses a continuously-fed wire as the electrode. This makes MIG welding very fast and clean.

situations. The D.O.M. mild steel tubing will give some and return to original position race after race. It is more ductile.

The 4130 material needs to be heat treated after welding to normalize the metal so it does not have an area that is weak or brittle. Usually the weak point will be the area right where the heat concentration stopped while welding. Also, the correct rod or wire must be used if you want to weld 4130 properly. 4130 has a lot more carbon content in it and therefore must use a rod with more carbon.

The 4130 material is stronger and more expensive, but not necessarily the best to use for stock car chassis construction. The mild steel D.O.M. tubing is best, provided the chassis structure is designed properly. This may not be the opinion of every chassis builder and racer, but GRT has currently produced approximately 600 late model cars with 95% being built with mild steel tubing, so the record speaks for itself.

Welding Methods

Welding the mild steel tubing is relatively a simple procedure if you are using a MIG (metal inert gas) welder. We weld each and every chassis with a MIG and the welds are smooth and flow-in well. The MIG creates a good strong weld and they are very adjustable for different types of material and wall thickness. What sets MIG welding apart from other types is that the electrode is a continuously-fed wire. This makes its weld rate (the amount of inches of weld laid down per hour) very fast, and it is very clean — no slag is left behind.

TIG (tungsten inert gas) welding — also called heliarc — is something that produces probably the nicest and strongest weld, but it is a lot more time consuming and difficult. We use the TIG to do all of our aluminum welding and some of our fabricated chassis components, but it is not necessary to use on general chassis construction. TIG is more suited for welding lighter materials like aluminum, and thinner pieces of steel. Welding steel tubing like that used for race car construction is a very slow process with TIG.

The TIG welder is actually better to use then a MIG because you can weld the tubing together in a smaller area. In other words, the weld or bead is not nearly as wide. The one problem is that the fitting process must be precise for TIG welding. When the fit is close, the TIG welding process is neat and small, but all of this takes lots of time and work. Then if you do it the correct way you must heat treat the welded region. So as you can see, this system is not practical for the weekend racer or even a chassis manufacturer. Another disadvantage is the higher cost of the machine. And, TIG welding requires a more experienced welder.

Arc welders can be used for welding on cages and frames, but again it is a more difficult procedure and is definitely harder to control than MIG. You must use a rod and it is continually being used up, forcing you to change it often. There is also flux that is built up around the head that has to be chopped off. This type of welder can be used, but it would not be worth the trouble for race car building on a regular basis. We don't recommend arc welding because of the size of metal you will be working with and the small areas that are involved in working on cages and frames.

Some of the more portable welders that are small, like a mini-MIG, work good, but sometimes they will not produce enough heat and don't penetrate well. So be careful about using a small MIG for this reason.

Welding The Chassis Together

The chassis is attached to the jig in several different locations to prevent it from moving and warping. We all know that welding pulls and warps the tubing

The proper notching and fitting of all tubing is an important element for improving the ease and quality of welds. The fit of the tube intersection at the left is very poor and should never be used on a race car chassis. All tubing should fit like the one on the right.

being welded, caused by the welding heat. What we have found to work best is to weld the chassis in the jig approximately 40 to 50% complete. Then we remove all bolt-on jig fixtures and remove the chassis from the main jig. The chassis is then welded the rest of the way on the floor. We have tried to do this many ways including nearly welding the chassis all the way up in the jig. What happens when we completely weld the chassis in the jig is that the heat pulls the chassis so much that it starts distorting the jig mounting parts and puts everything in a bind. The 50% welding in the jig and 50% welding on the floor works well because it doesn't seem to get things so hot at one time. We tested this procedure and put the chassis back in the jig after completion, and found that all jig points lined back up approximately 99%. Again this may not be the opinion of others but it works well for us.

The Mitler Brothers Ultimate Tubing Notcher is used primarily for high production work.

Another point about keeping heat and warpage at a minimum is the proper notching and fitting of all the tubing. Taking time to properly cut and notch tubing makes the welding job so much easier. The slag or excess material left on a freshly cut piece of tubing needs to be removed, and the fit must be checked before final installation of the tube. If this is done properly, the welding is so much smoother and quicker, and therefore making a nicer and cooler weld.

Tubing Notching

Notching tubing can be simplified if you have access to some kind of tubing notcher. We use a hydraulic notcher that is quick and easy for most all straight notches. For tubing that must be notched on an angle, we use the Ultimate Tubing Notcher built by Mitler Brothers. This really simplifies the process and once the angle is found you can mass produce the same notch as many times as necessary. These type of notchers don't leave any extra material around the notch so grinding or filing is at a minimum.

The Williams Lowbuck tubing notcher.

If you are just starting and are looking for a nice economical tubing notcher, Williams Lowbuck Tools makes the tool you need.

One other consideration of tubing notching is that the depth of the notch must be checked so when you cut your tube to length, you know how much to add to the measurement for the original cut. For example, if you have a 26-inch in-between measurement on 1.5-inch OD round tubing and the notcher jaws

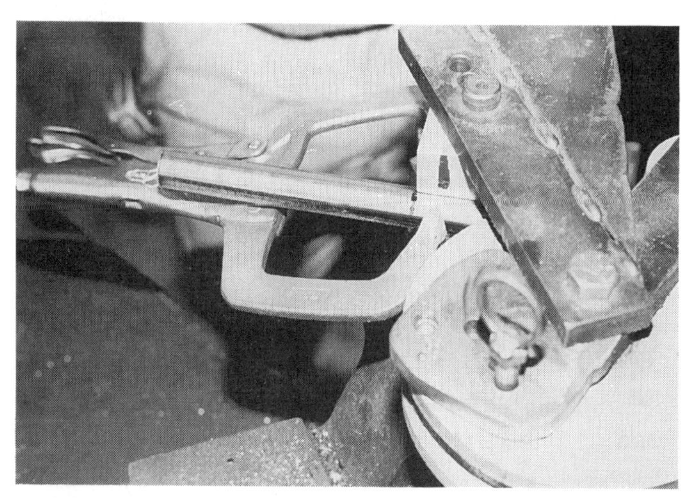

Clamp smaller I.D. tubing to prevent it from slipping as it is bent.

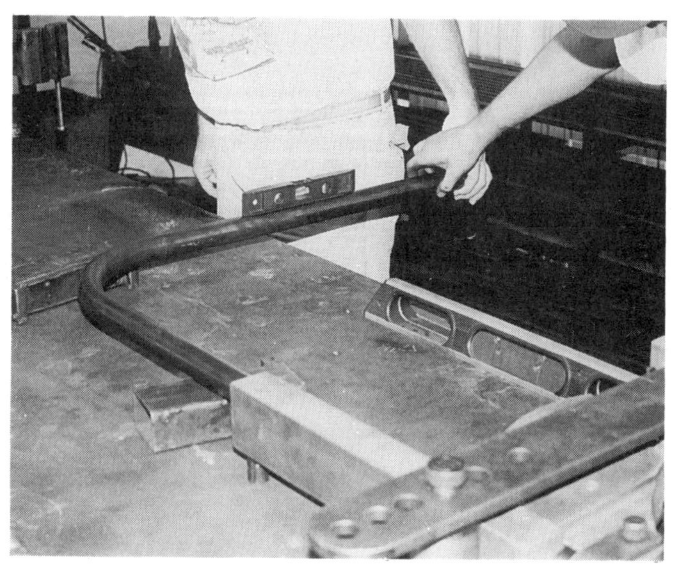

Use levels when bending tubing to make sure everything is flat and level.

remove 5/8-inch of material, you must add 1.5 inches extra to the measurement.

To calculate that, add 1.125" of extra material so if the notcher takes out 1.25" you will have .0625" slack on each side and this will enable the piece of tubing to fit nicely without any force. For example, 26" + 1.125" makes 27.125". Now you notch out about .625" on each end. This makes 1.25" total. You will have .125" slack total and you split the difference and it leaves .0625" on each side.

If you cut the tube just a small amount on the short side of the measurement (.0625"), it usually fits closer and doesn't require any force when installing the tube.

Bending Tubing

Notching tubing is one thing but bending is another. Bending tubing is a process that is simplified greatly by acquiring a good bender from a company such as Mitler Brothers Tools. They have formulas and instructions that make bending tubing easy.

Other tools that make the tubing bender much easier to use are the angle finder and a Smart Level. The angle finder allows you to determine the degree of the bend so you can have a guide to go by when starting to bend your tubing. Once you have pre-bent the tube and determine that it is the correct angle, then you can mark a spot on the bender itself as a reference point to perform the same bend the next time. Also, when using a tubing bender, meas-

Use a Smart Level to locate the next bend on a multi-bend tube application.

urements taken from where one bend stops to where another bend begins is an easy way to repeat bends. As always, keeping records of each and every bend or reference point on the bender simplifies the process.

Gussets & Bracing

As the chassis continues on in construction, there must be a certain amount of gussets and bracing installed in high stress areas. Gussets are always put in door bar areas where driver safety is a factor, and also in the top roof area. This really strengthens the area around the gussets. In case a severe impact or a shear type hit is made, the gussets keep the roll cage tubing from ripping apart. They do this by spreading the load on the tubing over a much wider area. Gussets add a lot of strength for such a small area and so little weight.

When installing the gussets in the door bars, you have to consider the door bar design. Remember that door bars which are built in a triangular manner from the inside of the car out seem to be much more crush resistant. Some door bar designs are a straight up and down design and won't resist side impact as well. In either design, gussets help strengthen the door bar area, but the triangular design of door bars helps add more protection for the driver.

A lot of racers put some kind of plate in the door bars to keep parts of other cars from piercing into the driver area (a lot of sanctioning bodies require this

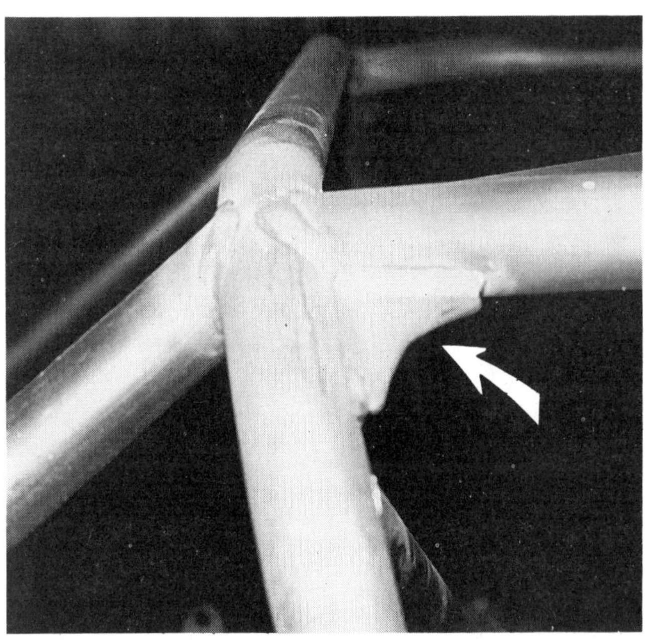
Gussets add strength to tubing intersections by spreading the load on the tubing over a much larger area.

Triangular shaped door bars add moe potection for the driver.

too). This is a good safety feature. 1/8 to 1/4-inch thick plate will be sufficient to provide protection. You can attach this by tack-welding the plate to the door bars, or by making a bolt-on application. I have seen 1/8-inch thick aluminum plate used before, but

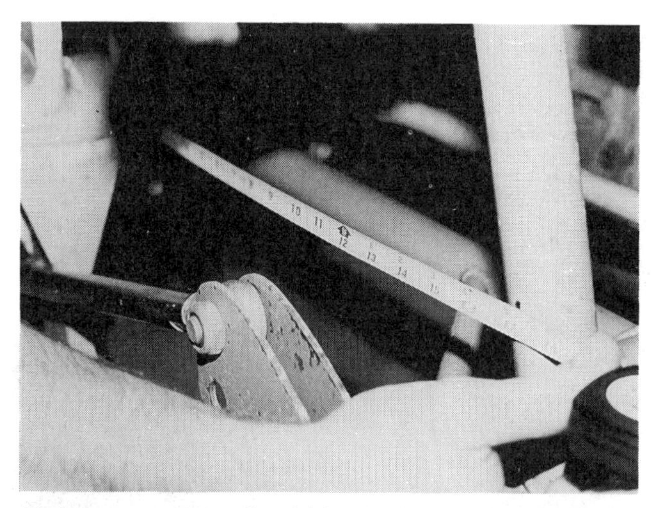

Reference points should be built into the chassis to check for chassis squareness and to square the rear end.

this is not as strong as steel. Remember that when you are adding the steel plate on the left side door bars, you are also adding to the desired left side weight.

Sometimes brackets can act as a gusset or brace in the chassis along with doing their job as a component attachment. To accomplish this, your chassis design has to be thoroughly thought-out.

Use a diamond-shaped flat plate gusset over a butt-welded joint on square or rectangular tubing. This improves the strength of the welded joint and spreads the loads over a wider area.

Chassis Reference Points

Chassis reference points are locating points on the chassis that the racer can always go back to for checking the squareness of the chassis or the rearend location. These are real helpful in case of an accident or crash and let you keep records of squareness on the frame. Check with your chassis builder for the reference points which are designed into your chassis, or you can add your own. Just make sure if you make reference points that they are square.

On our cars we square everything off of the right side main lower frame rail. This frame rail is the only one in the car that is 90 degrees to the rear axle. We square the rear end 90 degrees to this frame rail, having the rear end at proper ride height. Then we measure to a point on the frame straight in front of the rear end housing and center punch a mark. This

Underslung frame rails allows the use of ride height blocks to set the proper ride height.

measurement will always serve as a reference point for checking squareness of the rear.

One other advantage of our car is that we have underslung frame rails and this allows you to use ride height blocks in between the rear end tubes and the frame. These blocks set the proper ride height distance between the rear end housing tubes and the lower frame rails. While the blocks are in place, you can check all of your square points while the car is on jackstands. Also, our four bar brackets are square and if you put your radius rods at the specified length then you know your rear end is square. If someone crashes and they want to know if the frame is damaged, we give them measurements to check at different points on the chassis.

Fuel Cell Mounting

The fuel cell mounting in a dirt track race car should be as close to the rearend as possible and as high up in the chassis as possible. Normally the fuel cell mounts to the left side of the chassis. The fuel cell mounts need to be strong and made so that the fuel cell straps can bolt to the mounts and the fuel cell with the same bolt. In other words, bolt the fuel cell can and straps to the chassis mount with the same bolt. Fuel cells play a roll in handling and if you can make an adjustable mount, it makes it convenient to use the fuel cell for tuning the chassis. We will discuss this technique later.

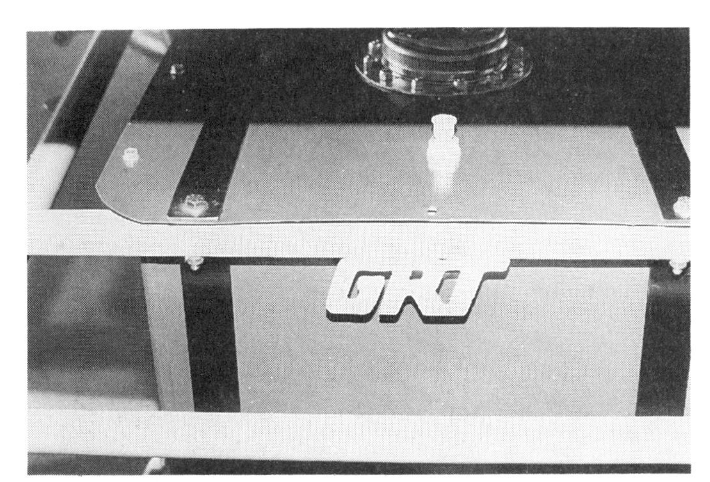

Fabricate the fuel cell mount so that it is easy to move the cell in order to help in tuning the chassis.

Nuts & Bolts

We have learned that grade 8 bolts are the best to use throughout the car. You don't need to have bolts breaking and possibly causing a crash or not finishing a race. The grade 8 bolt adds that safety factor. Another factor in bolting a car is to always use Nylock nuts. These will prevent a bolt from backing off or losing its tension on a race car. Another term for Nylock is self locking nut.

Nylock nuts have nylon installed in the outer edge of the nut just a bit smaller than the bolt diameter. When you completely screw the nut on the bolt, the nylon has to go over the threads. The bolt threads itself through the nylon while you install the nut. This prevents the nut from backing off due to vibration without using a wrench. One word of caution — don't use a Nylock self locking nut near extreme heat because it will melt the nylon and the nut will back off.

Remember that racing vibrations and impacts can easily cause bolts to lose their torque. Nylocks are a must, but if you can't use Nylocks, use Loctite on bolts.

Rod End Bearings

Late model dirt cars need to use good quality rod end bearings. We use aluminum rod ends everywhere on the car. They are lighter and do not wear as fast as a standard steel rod end, therefore they don't get as much slack in them and they last longer. The aluminum rod ends have a hard end steel ball and inserts that reduces wear. There are different grades of rod ends and we suggest you use a well known name brand. Check with the manufacturers or your parts supplier to find out specifications on rod ends, whether they be steel or aluminum.

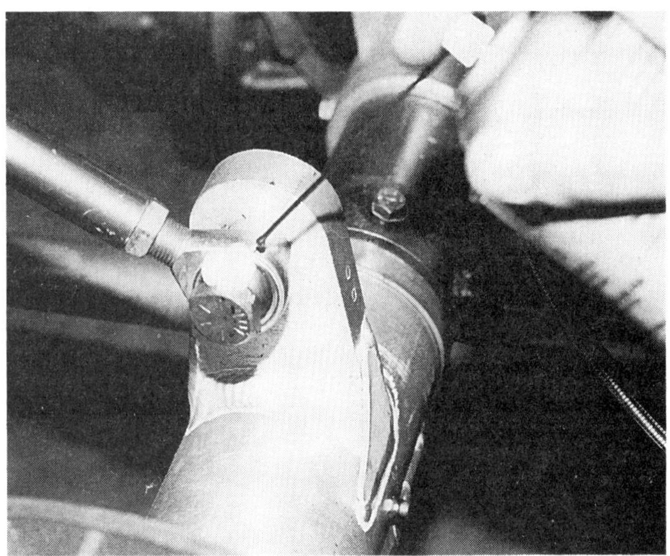

Always lubricate rod end bearings between races to prevent bearing wear and suspension bind.

Steel rod ends are fine but they require more maintenance due to the slack they get between the ball and the body.

In most applications, 5/8-inch rod ends are required for suspension and steering mounts, and are good for trailing arms and tie rods. 3/4-inch rod ends should be used on Panhard bars and steering supports. Rod end bearings used in shifting rods, power steering rods, or the throttle linkage should be standard 3/8-inch or 5/16-inch.

Aircraft quality rod ends are by far the best rod end. They are much stronger and have closer and longer lasting tolerances. For the serious racer with a larger budget, the aircraft rod ends are the ticket. In general we have found that a standard rod end is fine, but they do get slack in them faster on dirt track cars because of the continual dirt and sand. We always lubricate between races because this is a big factor in preventing suspension bind. Also, make sure rod ends are aligned square with each other on radius rods, and make sure that the rod end is as close to 90 degrees to the attaching point as possible to prevent the rod end from binding or breaking during movement.

Chapter
3

Mounts, Brackets, Interior, Pedals

Fabricating Brackets

When fabricating brackets you must first have a pattern to go by. You can use a piece of .040" thick aluminum or some similar material to build your pattern with relative ease. Then you can pre-fit your pattern and trim with snips, if necessary, to adjust the pattern for the best possible fit. After you have pre-fitted your bracket pattern, you can transfer it to the actual material that you make the bracket out of.

There are several different ways to cut and prepare the bracket you are making. A metal cutting band saw is a good way to rough cut your bracket. After you cut the bracket you form the edges of it with a small hand grinder and then finish it with a belt sander. Some brackets require a 45-degree or 90-degree brake or bend. This can be performed with a plate brake made specifically for bending heavier gauge metal.

Drilling holes for a bracket usually is done by locating the hole with a center punch and drilling the hole with a smaller bit to provide a pilot hole. This makes drilling a larger hole easier to locate for accuracy. If you are just an occasional fabricator or like working with your hands, these are basics to building brackets. If you are going to need several brackets, after you have the exact pattern of what you want, it is best to let a machinery business build your brackets in quantity. They can do this with precision equipment and each part will be exact.

Welding Brackets In Place

After you have fabricated the bracket you need, you must weld the bracket in place. This can be done by using a jig or placing the bracket by hand.

The best way to make sure brackets are square and true is to fabricate them in a jig.

Either way, make sure the bracket is square and true. Tack weld the bracket in place before completely welding and check for the fit of components to be installed to the bracket. Make sure the component to be attached clears everything around it. If everything looks fine, make a good solid weld for a nicely installed bracket.

Bracket Material Sizes

Bracket material used on most applications is 1/8-inch thick steel plate. This works well in most cases where there is not a lot of severe pressure on the bracket (for instance body tabs, radiator mounts, roof tabs, brake brackets, etc.). The 1/8-inch plate also works well for trailing arm mounts or radius rod mounts and the front Panhard bar mount as long as

The forward end of this trailing arm is captured in a double shear bracket. The rod end link to the birdcage is mounted in single shear.

3/4" x 3/4" .065" wall square tubing is used to mount all body components — front hood and nose mounts, rear hood and fender, rear deck and quarter panels.

they are designed as double shear brackets. Double shear brackets have bracket material on both sides of the attaching component (such as a rod end). This adds considerably to the strength of the assembly because the shear force is spread over the area of two brackets instead of just one. In single shear, the attaching components is just bolted to the side of a single bracket.

1/8-inch plate also works well on shock mounts, motor mounts bracing, bolt-on bumper plates and dry sump tank mounts. Fuel cell brackets, steering shaft brackets, upper seat mounts and motor plate tabs are also usually made of 1/8-inch thick material.

3/16-inch to 1/4-inch thick material works well for seat belt tabs or any type of mount where there is considerably more load on a bracket and you don't use the double shear method.

Some various sizes of square tubing are also used in mounting bodies. Transmission mounts will use 3/4" x 3/4" .065" wall square tubing. This same material is used to mount all body components — front hood and nose mounts, rear hood and fender, rear deck and quarter panels.

We use 1" x 1/2" x .083" wall rectangular tubing for upper a-arm chassis mounts, 3/4" x 1.5" x .083" wall rectangular tubing for lower a-arm chassis mounts, 1" x 1" x .083" wall for fuel cell mounts and the front bay cross member brace, and 1/2" x 1/2" x .065" wall for side body brace adjustment and hood pin mounts.

1" x 1" x .083" wall square tubing is used to fabricate the front bay cross member brace. Note the strong dual purpose bracket the cross member brace mounts into.

Using square tubing for brackets is usually stronger due to the fact that you can drill a hole all the way through both sides of the material and weld an insert as a bushing on two different sides. This gives the mount a lot of twist resistance.

Square tubing is used for the upper A-arm mounting bracket because it is easier to drill and weld in the threaded boss for mounting the A-arm. The square tubing also has a higher point of failure in case of side impact.

Bracket Fabricating Tips

The thickness of the tubing that you mount a bracket to needs to be close to the same thickness of the bracket or else it will rip out of the tube that is welded to. Another tip for preventing bracket failure is to have the bracket or tab made with breaks or bends in the bracket itself. For example, take a flat plate bracket and put a 90-degree bend on each end of the bracket for support. And, any place that you can, be sure to use double shear brackets. A double shear bracket is by far better than a single shear bracket.

Clamp-On Brackets

Another type of bracket is the clamp-on type of bracket. This particular bracket is very practical and convenient because you can mount the clamp anywhere on the chassis, and they are very strong and light. Some applications for the clamp-on bracket are ballast, shocks, body parts, dry sump tanks, and just about any other component you want to mount to the chassis. They make the clamp-on bracket in a variety of sizes for round and square tubing.

The flared-out bend on this straddle mount bracket adds tremendously to the strength of the bracket.

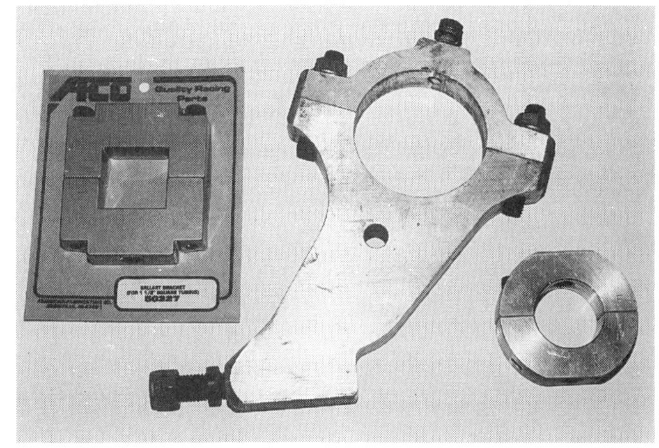

Clamp-on brackets provide a very practical and convenient way of attaching components to the chassis.

Interior Sheet Metal & Body Work

Another area of fabrication is the sheet metal and body work. .040"-thick aluminum sheets in baked-on pre-finished colors are by far the best type of material to use. This material is strong, lightweight, and very easy to work with. It is already painted and has an excellent repairability factor.

This same type of material is available in a .032" thickness as well. The .032" metal is somewhat lighter and not quite as strong, however. The main thing about .032" material is that it is lighter weight (a 4 'x 10' sheet of .040 is approximately six pounds heavier than a 4' x 10' sheet of .032" aluminum), but it will not withstand the abuse of dirt track racing. So unless it is a special application, I suggest staying with the .040" thickness.

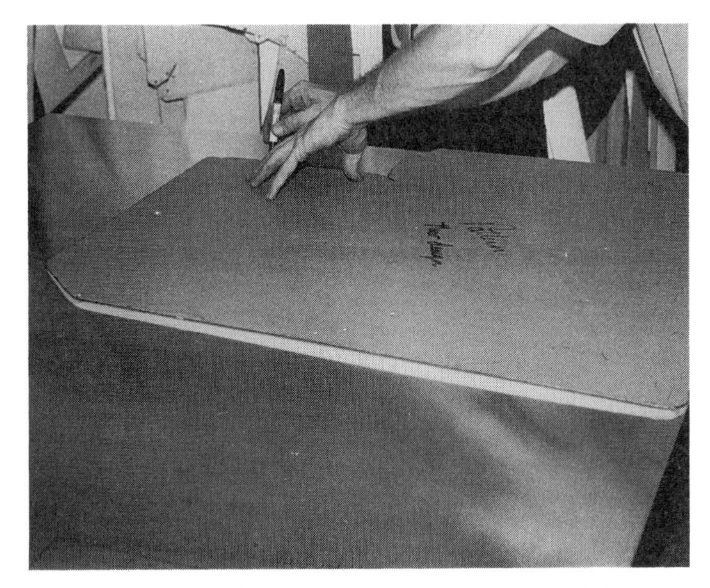

Use a cardboard pattern to lay out the shapes to be cut from aluminum panel stock.

Good interior sheetmetal installation starts with having the first piece installed properly.Make sure there are points to square off of. Then use Clecos or C-clamps to hold each piece in place while checking for correct fit or starting the next piece.

Fabricating Aluminum Sheet Metal

The most common type aluminum is the 3003 #14 for race car interiors and bodies. Designing your fabricated aluminum pieces starts with knowing what you want to build and which rules are enforced as far as dimensions and styles. When you are trying to fabricate a piece of aluminum, the best way to build a pattern is by using cardboard sheeting. You can cut and form this with relative ease. Once you have determined that your cardboard pattern is exact, lay it out on the material that you desire to use. You might want to take into consideration that if you are going to reuse the pattern you are building, then make a permanent pattern out of aluminum sheet metal.

You can make a pattern rack to hang all of your patterns that you make. This will simplify fabricating and sheet metal work tremendously when you can go back to use a pattern. For instance we have patterns for wheel wells, quarters, doors, fenders, nose pieces, hoods, and all interior pieces. Basically everything you build on a race car needs a pattern or measurement to go by.

Once you have your patterns and you are ready to start assembly of your project, there are several different processes to go through for installing the interior and bodies. You first must have the proper tools such as vise grip clamps, Clecos, drills, rivets, measuring tape, and a long piece of straight edge material. When you first start your job, you must fit the first piece and make sure there are points to square off of. Actually having the first piece installed properly will make the job much easier. Use Clecos or some type of C-clamp to hold each piece in place while you check for correct fit or to start your next piece.

Once you have the first portion of your sheet metal parts going together and you check all your fits and angles, then you are ready to start riveting together for the final assembly. Rivets should be put in approximately 3 inches apart on most interior cockpits and spreading out to as much as 5 to 6 inches apart on flat decking panels and body mounting.

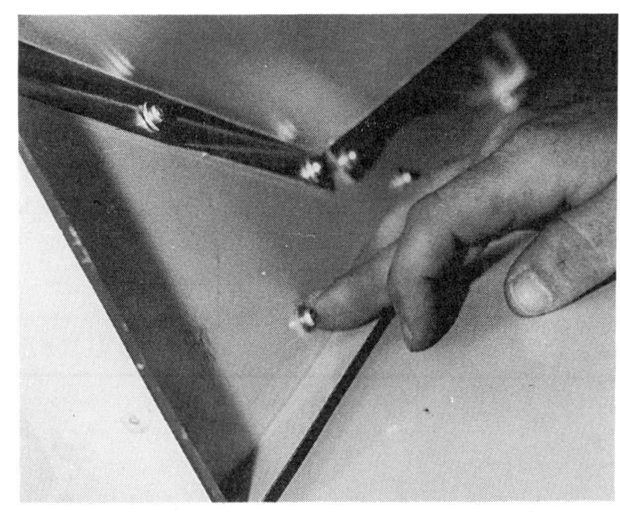

Using the aluminum mandrel and rivets makes repair jobs much easier when drilling the body panels apart for repair work.

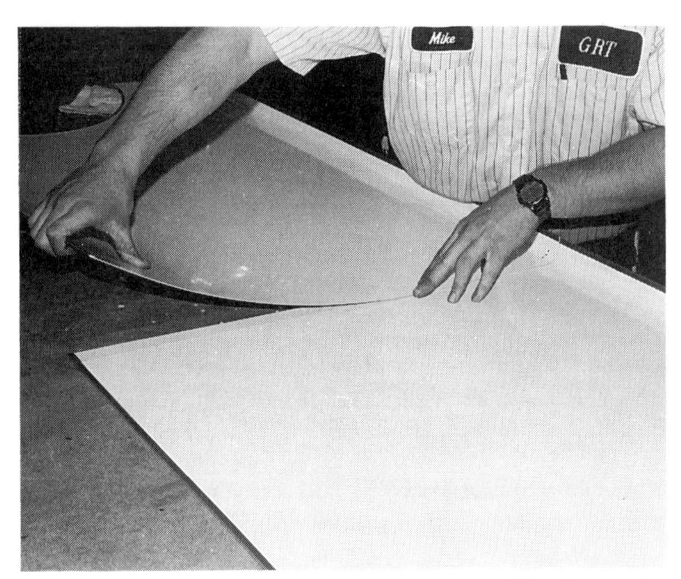

Using the two-sided 3M VHB tape to attach body panels together.

Also keep in mind that every time a hole is drilled, make sure that you debur or clean the area around the hole. This lets the material squeeze tightly together and your backup washer will fit tightly next to your panel when you pull the rivet up. Anywhere that you use an air rivet gun, the rivet will fit and pull up tighter for a neater and longer lasting job.

Using the aluminum mandrel and rivets makes repair jobs much easier when you drill the body panels apart for repair work. If you use steel mandrels it is extremely difficult to drill out, and it uses drill bits up. Steel mandrel rivets do pull up tighter but dirt late models will hold and stay together fine with aluminum mandrel rivets if the job is done right.

Remember to always check your parts as you go by using clamps or Clecos to hold things in place. It will save you a tremendous amount of time and assures you the job won't have to be done twice.

On some of our body panels we are putting together now, we are using a 3M two-sided tape for assembly. This works well in that the job looks cleaner and the decals that are used today go on much easier. The brand name of this tape is 3M Scotch VHB (very high bond), part #4930 95077 047N. This is an extremely high bond tape and is superior in quality.

Building Dash Panels

One area of sheet metal fabrication that is a little more difficult to build is dash panels. When laying

Put the gauge panel directly in front of the driver so he can glance at gauges without being distracted.

The prewired gauge panel is inserted into a fabricated aluminum panel. Notice how all the wires and lines are neatly run. Also note the rubber protection over the sheetmetal hole that the lines pass through.

out the dash panel you must first find the best location. We have found that putting the panel as close to and directly in front of the driver enables him to glance at gauges without distracting him from his concentration on the race track. After you have determined the location, the best thing to do is make a panel designed to let your gauge panel or gauges fit into (see photo).

We have found that building a preliminary panel just a bit larger than a gauge panel assembly simplifies the installation. First, center your panel into the aluminum panel you have fabricated. Then mark your holes similar to the way the gauges are shaped so the panel will set right down into the hole. Then

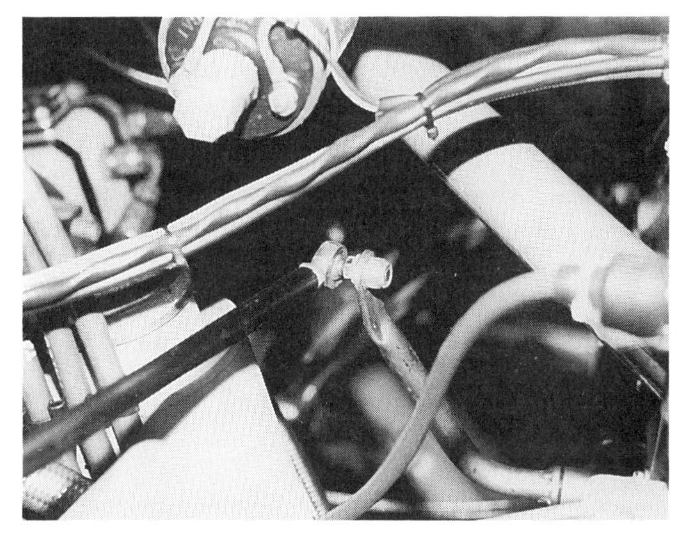

This throttle linkage is shaped as it is so that the throttle pedal connecting point is as low as possible and the carburetor connecting end is as high as possible. Make sure the linkage is free of any bind and there are no clearance problems. Always use double return springs for safety.

install with rivets in about 6 to 8 different locations around the gauge panel.

If you choose not to use a gauge assembly that is already pre-assembled, then you can install single gauges and warning lights into your panel. The pre-assembled gauge panels are the best way to go. They save you a lot of time in fabricating and assembling.

Throttle Linkages

The throttle linkage is a very important area that a lot of racers overlook. One of the biggest reasons this is so important is the quickness of the ratio. You want the throttle linkage to be as slow as possible. This makes the smoothness of throttle application a lot easier to control for the driver. You can obtain this by making sure that your connecting points are located properly. For instance, you want your throttle pedal connecting point to be as low as you can get it and the carburetor connecting end as high as you can get it. This slows down the ratio and makes throttle application as smooth as possible. This really plays a key role on extremely dry slick tracks and where extreme smoothness helps lap times. Remember the slower your ratio, the easier it is to control, and the faster the ratio the harder the throttle response is to control.

The throttle linkage should be free of any bind and double checked for clearance all the way through the throttle application. You don't want any interference from air cleaner bottoms, valve covers, etc.

Rod end throttle rods seem to work best and should be lubricated before each race. And, make sure you have properly installed throttle return springs. They will sometimes interfere with the stroke of the throttle shaft. This is a very important safety item. And, make sure your throttle linkage is adjusted properly to open the throttle all the way at the same time your accelerator hits the floor or throttle stop. This keeps you from putting the linkage and carb shaft in a bind. It also assures maximum performance because the butterflies in your carb are completely opened.

Brake & Clutch Pedals

In most dirt track racing, the clutch is not nearly as important as the brakes because you seldom use the clutch other than starting the car to roll, so we will discuss the brake first.

You must make the decision of which type of pedal assembly you want to use — floor mount or overhead hanging pedals. By far the hanging pedal is the best possible selection. It keeps the master cylinders up high and out of the way of dirt and mud or anything that might get thrown into the area of the master cylinders if they were low. Also, it is much easier to bleed the braking system because the brake fluid goes downhill and helps to bleed itself. Also, if you have a reverse mount pedal that has master cylinders on the inside of the car, this reduces heat

Exterior low-mounted master cylinders offer a couple of drawbacks. First, they will get full of dirt and mud. Second, they create problems in bleeding the system because the fluid flows uphill.

Make sure the balance bar adjuster cable is installed with no sharp bends or turns.

to the master cylinders because they are inside the interior and have a firewall insulating them from the headers. The nice feature about this type of set up is they are always in a clean area and therefore when you check the fluid there is no danger of getting dirt in your master cylinders.

One drawback to an inboard pedal is that sometimes clearances are a problem and you can't position the brake pedal properly. If this is a problem and you can't change the tubing in your chassis to accommodate this, use the swing mount pedals with the master cylinder on the outside. Either way it is a better brake pedal than the floor mount design.

A reverse mount pedal system places the master cylinders inside the cockpit, which reduces heat to the master cylinders.

Make sure that you allow enough area for the brake balance bar adjuster. You need to have plenty of room to run the adjuster cable and have no sharp bends or turns in it. This is so the adjuster can be turned easily. This is another area of a race car that, if overlooked, will create a lot of handling problems on all track conditions. So make sure the adjuster screws in and out properly and the driver has plenty of room to operate the lever.

When attaching the brake pedal to the chassis, make sure that it is securely mounted and has plenty of bracing. Any pedal mounting assembly which flexes will prevent the driver from creating proper braking pressure. His pedal effort will be used up in the flexing of the mount.

If you have the type of pedal where the master cylinder mounts on the outside, the firewall will act as a brace and help support the pedal. You should be able to check back into the fabrication and bracket subchapter for more information on bracket mounting and placement.

The Tel Tac digital tachometer. The gauges below it are brake pressure balance gauges.

Notice how each gauge has a corresponding warning light to the upper right of it.

Clutch Pedal & Linkage

The clutch pedal in today's racing is not a real critical point with the transmission we currently use. Basically any clutch pedal will do the job so long as you can "push her in and let her out". The reason is the clutch is only used when you start the car from a dead stop. You never use the clutch again because the car stays in one gear while racing. Therefore, the clutch pedal is just used for getting the car rolling. Usually the brake pedal that you use has a mate to it for a clutch and they work fine. The clutch linkage is a different story, however.

We prefer the hydraulic linkage over any mechanical type linkage due to the fact that most mechanical linkages bind up, require return springs, and need constant maintenance. We are not saying that mechanical linkages will not work, but the hydraulic is by far the best. Make sure the mounting location and the bracket being used on the throw out bearing linkage does not flex. Those type of linkages have a tendency to flex, so secure or brace it properly.

Gauges & Warning Lights

Gauges are a very important area for protecting the health of your racing engine. If you are going to have some kind of engine failure or you need to tune your engine to track size and conditions, then gauges and tachs are a must. You must first determine what type of gauge you want to use. There are several different manufacturers and two basic styles.

First let's talk about the liquid filled gauge. This gauge, in our opinion, has advantages and disadvantages. The main advantage is that the gauge needle is more stable and easier to read while under race conditions. One thing to consider is that while racing on most dirt tracks, the time you need to look at gauges and the time you need to look at the race track are very different. In general most tracks are short and you must concentrate on what you are doing. Reading gauges while racing is tough. Most of your gauge readings are done under cautions or after a race or during hot laps.

For instance most racers use a memory tach that will give the highest reading during hot laps or heat races. This will determine the gear ratio needed for optimum performance in the engine. So you will not need to look at the tach during racing conditions — just after you hot lap or during a caution. This same process usually goes for most other gauges like water temperature, fuel pressure, oil temperature, and oil pressure.

An experienced driver can pretty well glance at the gauges and see where the readings are and still concentrate on his driving. This is where warning lights come into effect (we will touch on this later). For a driver to be able to glance at gauges with very little distraction, the best place to install the gauges is directly in front of the driver's sight and up as high

Make sure all wiring and gauge lines are neatly tie-strapped and run through grommets or protected holes in the firewall or interior panels.

as possible. This will make glancing at gauges much easier, and the driver will not spend a lot of time searching for the gauge while racing. This is the main advantage to using a liquid filled gauge. It will stabilize the needle so when the driver does look at the gauges, it makes them easier to read.

Warning lights are just as important as any gauge to a driver. When you are racing and you catch a glimpse of one of these warning lights, you know immediately that there is a problem somewhere. The lights are a different color and it is easy for a driver to recognize that particular color and will be able to diagnose the problem immediately. Warning light colors are usually red light for oil pressure, white light for fuel pressure, yellow light for water temperature, etc.

In dirt track racing, liquid filled gauges are not a must, but they offer some advantages. A good quality gauge is all you need, but good warning lights are a must.

Installing gauges is very much simplified if you use the prefabricated gauge panels because all of the preliminary wiring and warning lights are already installed. Taking single gauges and wiring the lights up one at a time can be done, but it is not necessary or time-effective. If you purchase a prewired gauge panel, everything is already done for you except the actual installation of the panel itself. All you have to do is build some type of dash panel with a pre-cut a hole in it and install your gauges and lights. The best method of installation is using pop rivets to install the panel in place. This makes a very neat and clean installation. Make sure that all wiring and gauge lines are neatly tie-strapped and neatly run through grommets or protected holes in the firewall or interior panels.

Seats

Seats are a subject that racers must take seriously. There are so many factors that go into seats and their mounting. First of all, a seat should be as comfortable as your living room recliner but yet it needs to fit your body snugly. A driver must be comfortable to take care of business on the race track. You don't need to be sliding around in the seat or have any part of the seat gouging your ribs. So one of the first things to remember when choosing a seat is be particular about the seat and the fitting of the seat. There are many brands of seats out there that are quality seats, and some manufacturers will even custom build a seat to your specifications.

GRT recommends the aluminum seat over fiberglass or plastic. There really is no comparison in our

Racing seat attachments — lumbar support, head and shoulder rests, leg supports, and double rib cage wrapping — help support your body and increase safety.

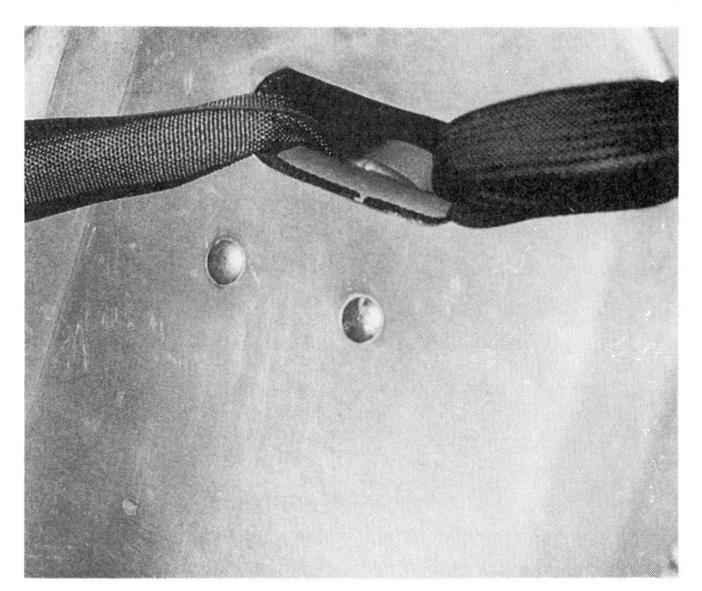

Use a seat mounting bolt with a large shoulder and a smooth head. This prevents pulling the bolt through the material, and the roundness of the head keeps the bolt from cutting or digging into the driver's back or legs.

opinion. The aluminum seat is by far the safest and sturdiest seat, so don't try to save dollars here — get the aluminum seat. The seat material thickness should be at least .090-inch. This is a minimum.

Most manufacturers also make an economy version of the aluminum seat. It is a universal seat that fits all types. This is my second choice for a seat and is by far better than the fiberglass or plastic seat.

Aluminum seats also have attachments — such as lumbar support, head and shoulder rests, leg supports, and double rib cage wrapping — that help support your body and increase safety.

Each aluminum seat manufacturer usually has a cover designed for the seat and this provides padding and support.

Some drivers don't like seat covers and race without them. In this case you really need a seat that is custom fitted so it does support the body extremely well. For instance, Bill Frye does not use a seat cover and he custom builds his own seats. He says he is much more comfortable and in control by getting that "feel by the seat of your pants" in a seat that has no cushion. This may be fine for him but if you can't build your own seat, purchase one from a manufacturer that builds quality seats and has a cushion designed for that particular seat. D & M and Butler-built are just some of the seat companies you can rely on for a good product.

One other feature that I like to see in a seat is the under-leg support which produces a slight bend in your knees. Again this is a driver preference, but it supports the bottom of your legs and you really gain extra support and control.

Using an aluminum seat also makes the mounting procedure easier and safer. If the mounting procedure is not done properly, it will damage the seat and possibly injure the driver during an impact. This problem is compounded if the seat is made of fiberglass because it will break easier (fiberglass or plastic will flex easier also). Make sure the seat is well supported by some sort of seat mount and firmly bolted to the chassis. Use a bolt with a large shoulder and a smooth head. This prevents pulling the bolt through the material, and the roundness of the head keeps the bolt from cutting or digging into the driver's back or legs.

Some seat manufacturers recommend you build a frame to support the seat that is welded to the chassis. This is fine. Just remember that the seat needs to be firmly bolted in with support over a large area. We also recommend that you have a minimum of four mounting locations on the seat — two on the bottom and two on the top.

The best possible location to mount the seat is as far back in the car as is practical, and centered in the driver's compartment between the driveshaft tunnel and the door bars. You must make sure there is ample arm and head room so the driver doesn't contact anything. You don't want the seat to be too close to the door bars because the driver could injure his elbow or hand if he accidently hit the roll bar tubing. Make sure there is ample head clearance also. The driver's head should be a minimum of 4 inches away from any roll cage tube when he is tightly belted into the seat.

Another option for installing a seat is to tilt or lean the seat to brace against the G-force the car will produce during racing. This will also increase body support and help keep the driver concentrating on racing and not trying to hold himself in the seat.

We fabricate seat mounting brackets from 1/8" steel plate for the rear seat supports, and use 3/4" x 3/4" tubing formed to the seat for the lower mounts. You can also mount the bottom of the seat with the same bolts which fasten the lower seat belt mounts. We mount the lap belts sandwiched in between the lower portion of the seat and the mounting bracket.

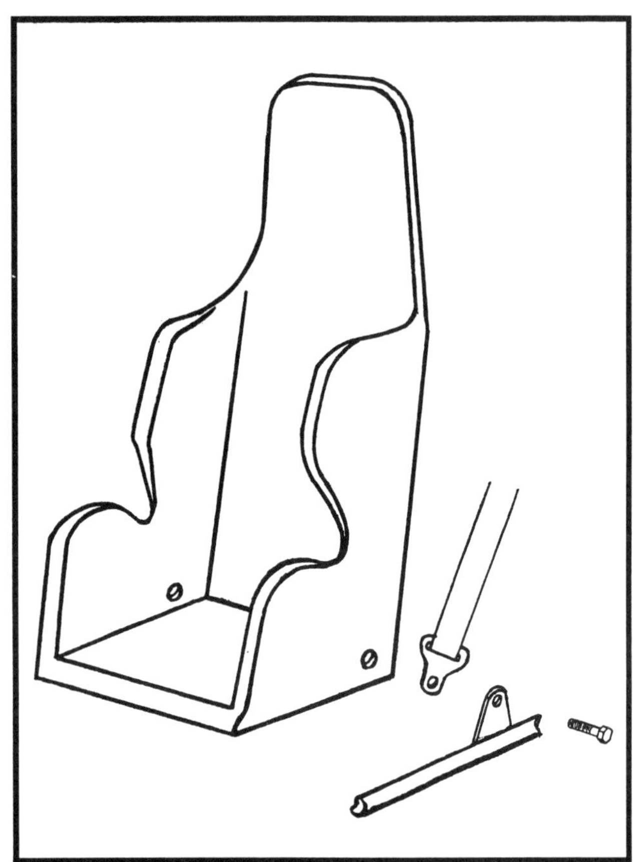

We mount the bottom of the seat with the same bolts which fasten the lower seat belt mounts. The lap belts are sandwiched between the lower portion of the seat and the mounting bracket.

Seat Belt & Shoulder Harness Installation

We strongly suggest that you use 1/2-inch grade 8 bolts to mount belts and harnesses. Shoulder harnesses must be run through the hole provided in the back of the seat rather than being routed around the edges of the seat. Make sure that there is no interference or sharp edges anywhere near the shoulder harness that could chafe or cut the belts.

The mounting location for shoulder harnesses should be directly behind the driver's shoulder blades (see diagram). There should be no bend or misalignment of the harnesses while in use. Lap belts need to be securely bolted to seat belt tabs or brackets welded to the chassis.

Some seat belts have a mount that bolts to the chassis and a latching type seat belt may be used. This enables the racer to take seat belts out when the car is being washed. If you do not take belts out during washing, they may lose their strength somewhat during a season of washing and drying. We suggest that no matter what kind of belt you use, replace them once a year for your own good. They do wear out from continual exposure to oil and grease, dirt and sunlight The best type of belts to use are 3-inch wide lap and shoulder harnesses with a 2-inch sub belt. Remember that all belts need to meet the SFI 16.1 standards.

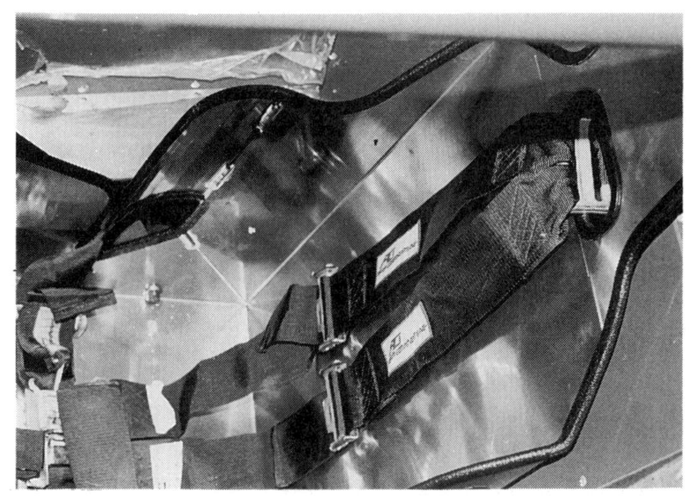

Make sure there are no sharp edges anywhere near the shoulder harness that could chafe or cut the belts.

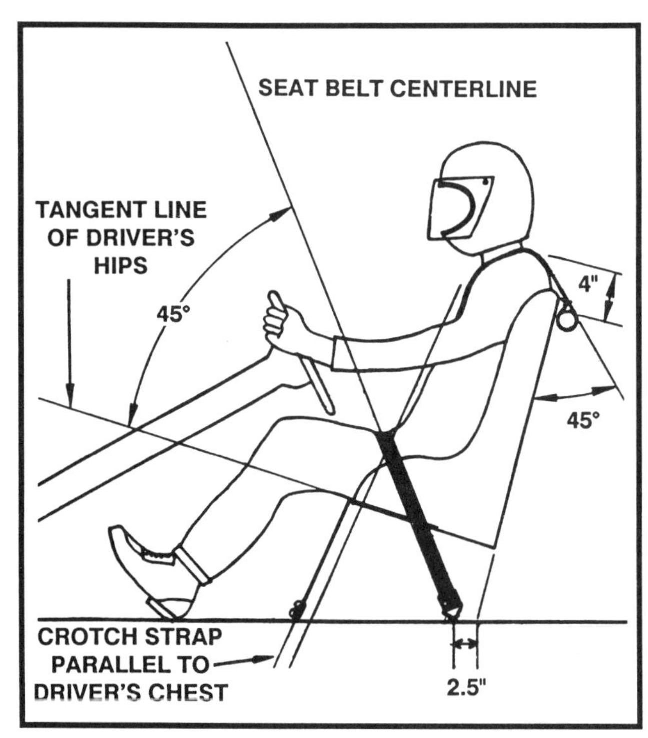

The correct mounting locations and angles for the driver restraint system.

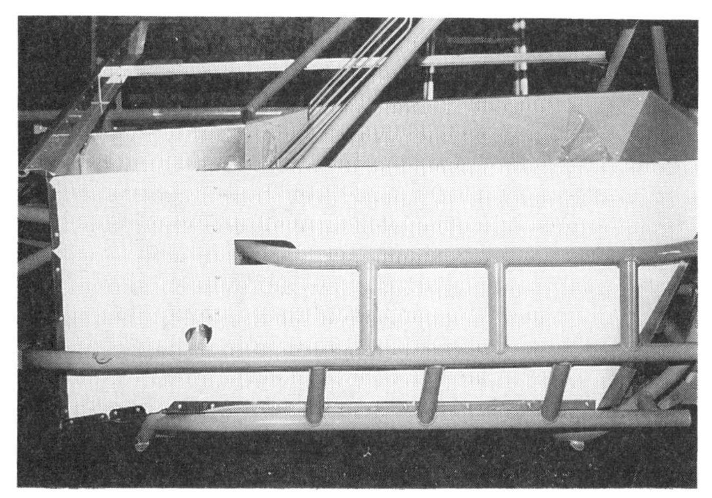

Enclosing the left side of the driver's compartment with aluminum sheeting creates a smooth crushable area that will assist in protecting the driver in case of impact.

When mounting the sub belt, it needs to attach to the lower bottom seat mount or a belt tab on the chassis just for the sub belt. Make sure the proper mounting angle is maintained (see diagram).

Some drivers prefer padding on their shoulder belts and there are a variety of safety options beyond the standard belts. If padding is used, make sure that the harnesses can still be cinched down tight so there is no slack in them at all. Make sure you have a quality set of belts and they are mounted properly and they meet the current safety standards.

Roll Bar Padding

Roll bar padding is a major safety item used to cushion a driver from impacts against roll cage tubing in case of an accident. The best type of roll bar padding is the high density foam type with the offset center hole. The usual installation procedure is with plastic tie straps or wire ties to hold the padding securely in place. The padding should be placed on any tubing around the driver that he could possibly come in contact with in the case of an impact. Remember that under severe impact, a driver's body and the belts which contain him will stretch, so make sure that padding is added to tubes further away from the driver.

In general the driver's compartment should be as clean as possible with no sharp or blunt objects near the driver. Anything that the driver can possibly hit or move against needs to be removed or relocated so injury will not occur. Make sure the driver has plenty of head room, leg room, hand and arm clearance. Use padding on any place that might inflict driver injury. Do not mount anything in the driver's compartment that might shake or jar loose in case of an accident or during a race.

One other area that contributes to driver safety is to enclose the left side of the driver's compartment with aluminum sheeting. This keeps arms and legs from possibly getting into the door bars and creates a smooth crushable area that will assist in protecting the driver in case of impact.

Window Nets

Most sanctioning organizations require the use of a window net. A window net will help keep the driver's arms and head in the car should his body stretch that far, and the net will also prevent debris from hitting the driver. The window net attachment to the car should be carefully considered so that it is positively retained (most organization's rule books have specific mounting details). Make sure the window net fits snugly when it is in place so that stretch in the material is tight enough to prevent the driver's head or arms from going outside the window opening plane during an impact. And, the net should be very easy for the driver to open and remove from inside the car, even if there has been some damage to the driver's side of the car.

Driver Safety

The driver also plays a role in creating a safety factor for himself. The driver should always wear fire retardant gloves, shoes, underwear and driving suit. There are several different types and qualities of material used for driver protective clothing, and the driver should research the various types and make himself familiar with all of them before making a purchase. And, a good quality full face helmet is a must. Don't cheat in this area. It could mean your life, or being injured for a long time.

Chapter

4

Wiring & Plumbing

The Electrical System

First we will start with the battery. This is a very important part of the system because it provides the initial starting power and in the case of a regular ignition system (any electronic ignition where a magneto is not used), it maintains the fire to the entire system. But in most late model dirt cars, you won't see many electronic distributors — just magnetic type ignitions. In this case the battery doesn't play the same part as it would in the electronic ignition system. Regardless, you need a quality high cold cranking amps, lightweight battery such as an Interstate MT-26 Mega-Tron. This battery is 8" wide, 7" deep, and 7" tall. As you can see it is very small and lightweight. It weighs only thirty pounds. It has a cranking capacity of 650 amps.

With the size and weight of this battery, it is a very practical and functional power source. You can also place it in the chassis more conveniently due to its size. There are other batteries that are similar in size, shape, and power that are good — this is just the battery of choice for GRT. The smaller dry cell batteries have been tried and we have not had good success with them. They seem to not have the juice required to turn today's high torque starters.

The size of the battery cable used is very important in determining how a race car starts. If the cable is too small, then no matter what size of battery is used, improper starting may result. We tried to get away with using a #4 battery cable for many years, but it is not large enough to carry the current it requires to start race cars. We suggest using a #2 size cable. This is like bolting in a whole extra battery.

If you use a magneto, the battery really does nothing except start the engine and provide power to the gauges for lights. If you use a standard or electronic ignition, then keeping a well charged battery is a necessity.

Most short track race cars do not use an alternator. In order to keep a fresh, charged battery in the race car, charge the battery before each racing night. Under normal use, this will provide sufficient charging (provided you don't do a lot of extra cranking on the engine). Even if your distributor is an electronic type, the laps run on a normal Saturday night race won't drain a battery enough to affect the system. But, always keep a spare, well-charged battery with you at the race track.

Note the simple ground connection on this battery. It uses a piece of 1" x 1/8" aluminum which bolts from the battery to the chassis. No clamps or wires are there to come loose. It serves as a battery hold-down as well.

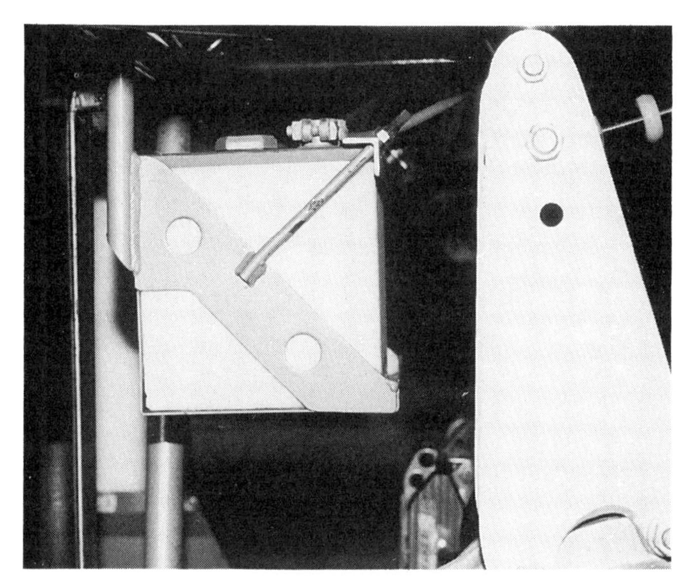

Battery placement should be on the left side of the car in the rear, near the rear axle, about the same height as the top of the fuel cell.

An electronic ignition system will drain the battery more severely, so you must keep an eye on the battery and cables more than you would with a magneto ignition. In general, no matter what kind of ignition you have make sure you have quality, properly sized cables and that the ends are put on with a good crimper. Get a good quality battery and keep a battery charger on the system between races, and you will have a functional system.

Battery Placement

Battery placement should be in the rear of the car around the rear axle, about the same height as the top of the fuel cell on the left side of the car. This battery placement helps to take maximum advantage of the weight of the battery for rear percentage and left side weight. Also, a battery that is placed higher in the chassis creates more overturning moment, which is a benefit on dirt cars for creating side bite. When a battery is placed this far back in the chassis, the size of the battery cable and good solid connections become even more critical.

Alternators

We do not suggest using an alternator. It is not a necessary item on late model dirt cars. With the number of laps run and the limited use of the battery capacity, using a fresh, fully charged battery provides sufficient electrical source capacity for the race car. Considering that an alternator is another item that requires maintenance, and it pulls horsepower away from the engine, we feel an alternator is not required in this type of racing.

Wiring The Car

There are several different methods of wiring a race car, but what we explain here is the most common method (see wiring diagram next page). It is how the wiring is run on GRT late models. Included is a diagram for reference.

The first step in a wiring system is having an ignition panel which includes switches to turn the distributor or magneto on and off, and a starter button that will turn the engine over with a push of the finger.

First, run a main hot wire from the hot side of the solenoid (which is getting its power from the battery) to the starter button terminal. For this wire we use a 16-gauge PVC-coated red wire. (All of the wiring we use is 16 gauge.) Then from the other side of the starter button terminal, we run a yellow wire to the solenoid of the starter. With another red wire we jump across from the starter button to the gauge switch. A green wire is connected from the gauge switch to the gauge panel for gauge lights and warning lights. Most gauge panels are already prewired, so you will not have to do any wiring there.

To wire the magneto switch, we use a yellow wire going from the terminal of the mag switch to the distributor hook up. The mag switch is just a two terminal switch that is either open or closed. When the switch is open, it grounds the magneto and kills the engine. When the switch is turned on, the mag functions. From the other side of the mag switch a white ground wire is run to a good chassis ground.

Depending on the type of magneto you have, it will either have an internal coil or an external coil. If it has an external coil, you must run a ground wire from the coil to a ground source, and run a blue wire from the magneto terminal to the external coil.

The last item that needs wiring is the tachometer. A black wire is run from the magneto to the tach, then a white ground wire is run from the tach to a chassis ground.

The main thing that you need to be concerned about when wiring the car is to make sure all of the wiring is done neatly and safely. For instance, make

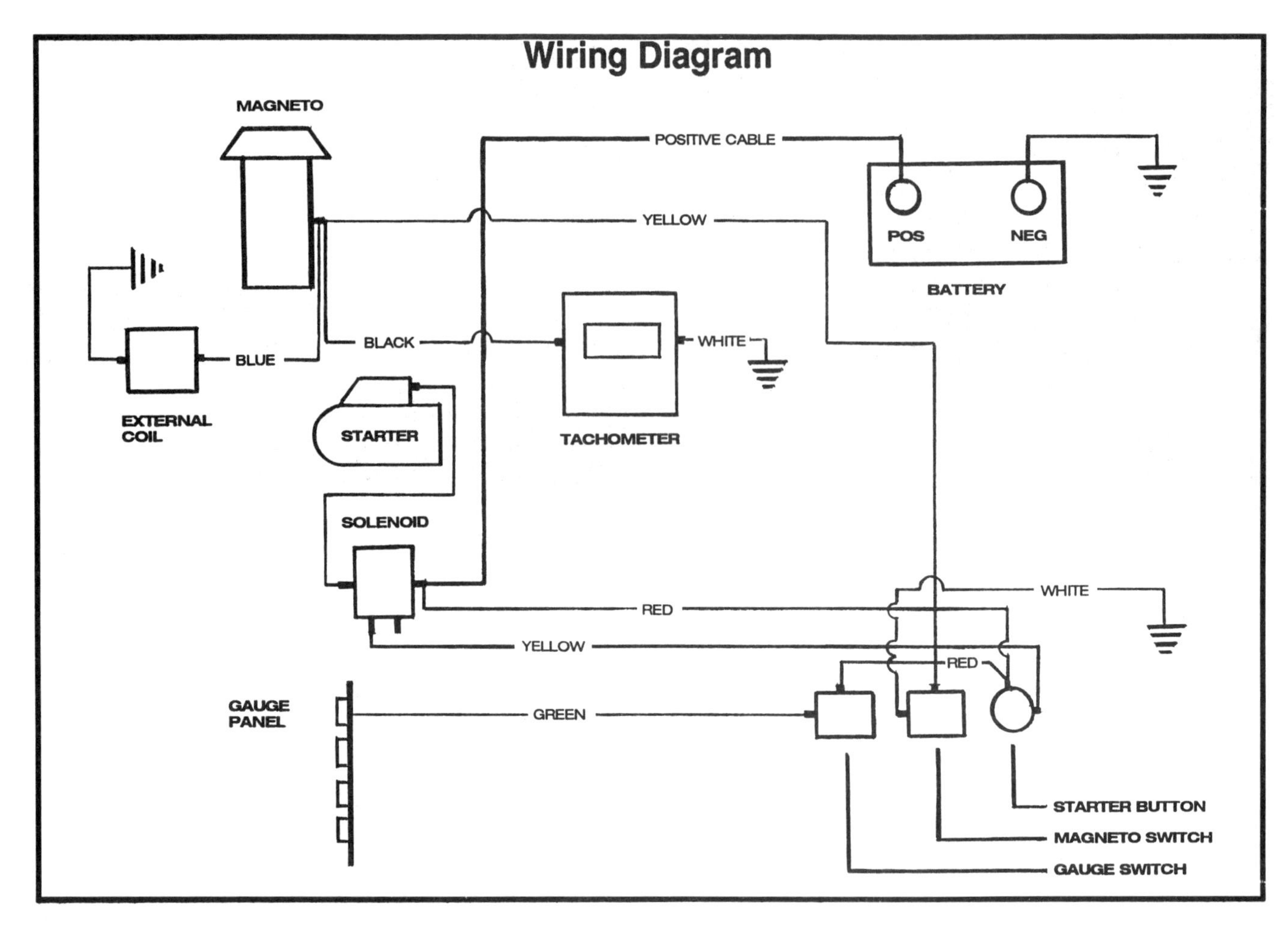

sure all wiring is done as straight as possible and there are no sharp objects or edges in the way that might cut or chafe the wire. Never route wiring near any high heat components. Run the wires together and use wire tie straps to hold wiring neatly in place. Anywhere that wires have to go through a firewall or interior panel, be sure to use rubber grommets to protect the wiring. Don't let any wires hang loose or droop down. They need to be run in line with interior panels or along roll cage tubing. Most electrical stores sell a stick-on wire mount for use on interior or body panels. It has a clip on the other side to attach wire ties. This allows you to wire tie your wiring job in a very neat manner.

Color coded wiring can be very helpful in a race car. Use different colored wires for each type of hook up, and keep records of which wire colors go where. In case a quick repair is needed, you know exactly where a certain colored wire goes.

All wire ends should use a ring type of connector so the wire can't accidentally fall off of a post or bolt. An open ended U-shaped lug could possibly work

All wiring should be as straight as possible with no sharp objects or edges in the way that might cut or chafe the wire. Use wire tie straps to hold wiring neatly in place. Be sure to use rubber grommets to protect the wiring where it goes through a panel.

loose due to vibration. Make sure you use insulated type wire ends and use shrink tubing wherever possible. The shrink tubing will tightly encase the wire/terminal end connection and will prevent a lot

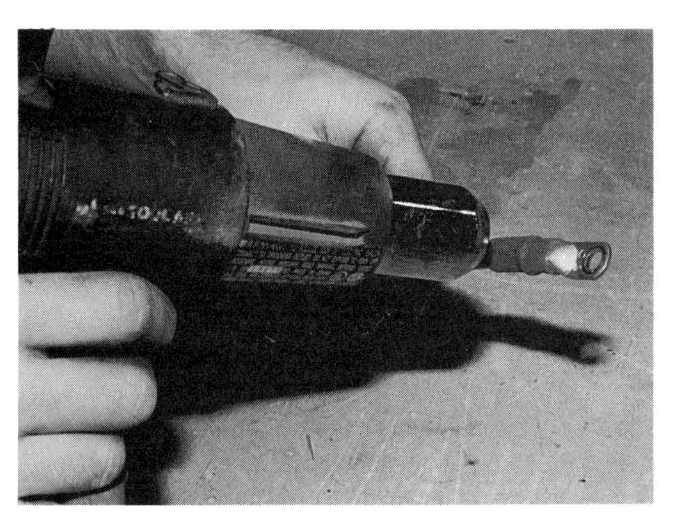

Shrink tubing being applied to a wire that has a closed ring connector end.

of problems. Make sure all grounds are grounded with a bolt and not a pop rivet, and be sure all grounds are properly tightened so they will not work loose.

Battery Disconnect Switch

An electrical kill switch, or master disconnect, should be installed to shut off all electrical systems in the car in case of an emergency. The battery disconnect switch should be in reach of the driver and a crewman who approaches the car from the outside. If there is an accident or emergency where the battery power must be cut off and the driver is not able to get to it, a crew member or track official should be able to reach the switch. The switch should be mounted so that the negative side of the battery is disconnected by the switch.

Plumbing The Car

There are several types of hoses and fittings required for a functional non-leaking plumbing system in a race car. The car requires oil lines, power steering lines, fuel lines, brake lines, gauge lines, clutch lines and water lines.

Most lines in a race car will be plumbed with stainless steel braided hose. While many people think (or assume) that all steel braided hoses are the same, there are many different types and grades. It is very important to know and understand the differences so that the proper type of hose can be selected for each application.

An electrical kill switch, or master disconnect, should be installed to shut off all electrical systems in the car in case of an emergency.

Oil Lines

Oil lines will usually range in the size of #10, #12, and #16. The hose size numbers refer to the inside diameter of the hose (see chart for complete listing). The hose ends need to be the same brand as the hose used and you need to make sure the ends are installed properly. Each fitting manufacturer includes an instruction sheet which explains the installation and hose cutting process in detail.

Normally straight, 45-degree and 90-degree fittings are all that you need to complete an installation job. Some tight fitting areas might require 120-degree or 180-degree fittings. These are handy where oil pumps or pans might be extremely close to the chassis or chassis components.

Stainless Steel Hose Specifications

Hose Size	Inside Diameter	Max. Operating Pressure
4	7/32	1,500
6	11/32	1,500
8	7/16	1,500
10	9/16	1,250
12	11/16	1,000
16	7/8	750
20	1 1/8	500
24	1 3/8	250

Fuel lines should be routed in the shortest possible route. The line should be tightly secured and away from heat sources.

The hoses we install are from Earl's Performance Products. For oil lines, we use their Perform-O-Flex hose. These hoses provide good flow and hose flexibility at temperatures from -40 to +300 degrees. They have a synthetic rubber inner liner with a partial coverage woven stainless braid embedded in the rubber liner. Then a high tensile strength stainless steel braid covers the outside and is mechanically bonded to the hose. This provides a maximum operating pressure up to 1,500 PSI, depending on the inside diameter of the hose used (see chart). The larger the diameter of a hose, the lower the pressure rating of it.

Fuel Lines

The fuel lines we use are Earl's Super Stock hose. This hose uses a synthetic rubber inner liner reinforced by a full coverage interior braided fabric and covered in a synthetic rubber material.

This type of hose has a 250 PSI maximum pressure rating. Normal sizes used are #6, #8, and #10. This hose is extremely easy to use and easy to route. It uses Super Stock hose end press-in fittings that are very easy to install. After you have established the required length of hose, cut it with snips or a sharp knife. Then simply press in the hose end fitting — no clamps or tools required. The fitting is held in place with an outer aluminum collar. These fittings are intended to be used only with the Super Stock hose, and no other type of hose end should be used on Super Stock hose. As a safety precaution, no push-on hose end should be used without some sort of clamping device. The Super Stock hose end is supplied with an aluminum sleeve which can be easily crimped with an Earl's Auto-crimp tool.

This type of fuel line is lightweight and very flexible and less expensive than using steel braided lines. We are not saying that the steel braided line is not good to use for fuel lines (because it is), but this type of rubber hose definitely has its advantages in being so easy to work with. (However, many sanctioning bodies prohibit the use of rubber oil and fuel lines — especially where they pass through the cockpit.) The fittings are machined from aluminum alloy and are anodized. They come in the 45-degree and 90-degree type for tight clearances and produce a neat installation. We recommend that the racer use a #10 line for the main supply and a #8 for the return line on an alcohol-fueled car. Most engine and carburetor builders recommend these sizes.

Fuel lines should be routed in the shortest possible route. We usually route the line down the right side upper snout bar to where it runs into an in-line filter near the rear wheel, and then down into the lower right hand corner of the fuel cell. This fitting in the lower right hand corner should be a bulkhead #10 tee. When you use a tee it enables you to drain fuel at any time without taking the main fuel line off. Always use anti-seize where aluminum fittings go together so you don't seize the threads. This is a good practice on any type of fittings.

Fuel Line Protection

The best way to protect fuel lines is by the way they are mounted. Make sure the line is not anywhere close to heat, sharp edges or close to the ground were it could be snagged during racing. Make sure the line is tied securely with tie straps or Adel clamps. When the fuel line gets near extreme heat, such as at the engine or headers, it is a good idea to insulate the line.

Power Steering Lines

We use Earl's Power Steering hose, which is a triple steel and fabric reinforced high pressure hose developed specifically for the high pressure of power

Brake flex lines run only from the chassis to the caliper to allow suspension movement and turning capabilities

Solid tubing brake lines should be attached to the frame or sheet metal with rubber-lined Adel clamps.

steering systems. They are extremely high pressure hoses that are rated at 2,250 PSI maximum operating pressure in the #6 size and 2,000 PSI in the #8 size. They will take extreme heat and are fairly flexible, but they have a minimum bend radius that is larger than other types of stainless steel braided hose.

The pressure lines in the power steering system are a #6 size, and the supply line uses a #10. Earl's developed special Power Steering hose ends which are made of steel for use with this hose in order to handle the high pressure application. They come in straights, 45-degree, and 90-degree. The fittings are installed the same as steel braided aluminum fittings.

It should be noted that power steering pumps and boxes are manufactured with different port sizes and configurations — JIC, male and female, inverted flare and even AN. Earl's supplies hose ends and adaptors for each configuration, but the different types will not mix.

Brake, Clutch And Gauge Lines

Earl's Performance Products manufactures Speed-Flex steel braided hose specifically intended for use in brake flex lines. It uses an extruded Teflon inner liner, with a tightly woven high tensile stainless steel outer braid to protect against brake line swelling during braking application. We use this same hose for clutch and gauge lines. All of these lines are run in the #4 hose size, which has a 2,000 PSI maximum operating pressure. Only Earl's Speed-Seal hose ends should be used with this hose.

Some chassis builders prefer to use #3 size brake flex lines in order to minimize line swell and the attendant pedal travel.

You should be aware that there are different grades of Teflon lined steel braided hose being manufactured. Some have a .030" wall thickness while others have a .040" wall thickness. The Earl's Speed-Flex hose is made with the thicker liner to assure maximum pressure resistance.

Brake flex lines run only from the chassis to the caliper to allow suspension movement and turning capabilities. The flex lines attach to steel brake lines that we route from the master cylinders.

For the solid, fixed brake lines in the system, use automotive double wall (Bundyweld) tubing with a 3/16-inch outside diameter. The tubing should be attached to the frame or sheet metal with Adel clamps. These are rubber cushioned tubing clamps which assure a tight fit and eliminate vibration. Secure the steel brake lines to the chassis in several places along its run with these clamps to prevent vibration fractures of the lines. All tee fittings along the steel tubing run should be secured to the chassis with a bracket.

When forming the hard brake lines, make sure you don't bend too tight and crimp the lines. Make bends precise and smooth. Use a Magic Marker to mark the center of the bend, then bend on the center of the mark. Use a brake line tubing bender (available at parts stores or tool stores) to make good, neat tubing bends. Bending by hand or over a mold will produce crimps in the tubing.

When the brake lines are cut to their proper length, use a double flare on the end. A single flare will crack and leak. And, always cut the steel tubing with a tubing cutter, not a hack saw.

Never route brake lines close to any moving parts that might cut the line or near any heat source such as headers, exhaust pipes and oil tanks.

Assembling Steel Braided Hose & Ends

Start the assembly process by cutting the steel braided hose to its proper length. Remember when measuring for an application to leave enough length to give the hose some slack. When cutting steel braided hose, begin by tightly taping the area to be cut with masking tape of duct tape. Masking tape is better. Cut it with a good radiac wheel or a fine tooth band saw or sharp hack saw blade. The tape will keep the stainless wire from fraying. Be sure the cut is made square with the hose to ensure a good fit.

After you remove the tape, trim any frays off with snips or cutters. Then put the threaded socket in a vice and install the hose into the socket (figure 1). Leave about 1/16 to 1/8-inch gap between the end of the hose and the bottom of the socket. Mark the hose at this point, then push the hose to the bottom of the socket. When the main fitting is screwed into the socket, the hose has a tendency to push out. By watching the mark you have made on the hose, you can tell if it is pushing the hose out as you tighten the fitting. If the hose moves out more than 1/16-inch, you will not have enough hose inside the fitting to maintain a good solid installation. If the hose backs out too much, start the assembly all over again.

Before you start this process be sure to lube the fitting and the inside of the hose with clean motor oil so it slides in easily (figure 2).

Push the socket and hose onto the nipple in order to start the threading on the nipple (figure 3). Turn it as much as you can by hand. To complete the threading, hold the socket in a vise and tighten the cutter threads until the socket is within .060-inch of bottoming on the nipple. This distance is important because if the socket is not sufficiently tightened on the nipple, leakage could result. Use a hose-end wrench when threading so that you don't make the mistake of using too much force with a larger wrench (figure 4).

Once the assembly is completed, clean it with solvent and blow lines clean with high air pressure. Brake flex line assemblies should be washed clean with new brake fluid. It is also a good idea to pressure check the completed hose and hose end assembly for leaks under pressure. Check it in its intended application, or use a pressure tester. Earl's sells pressure testing kits.

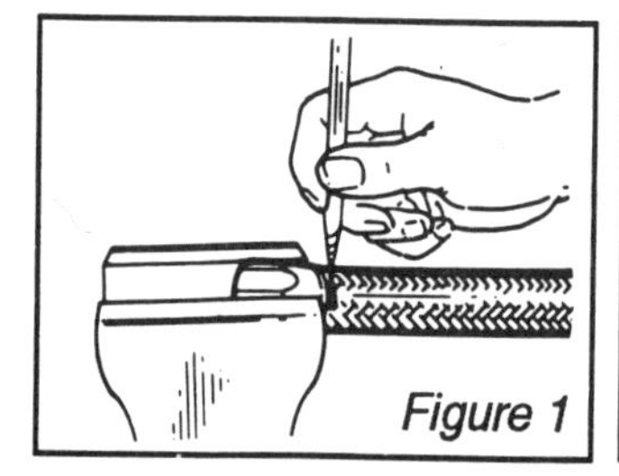
Figure 1

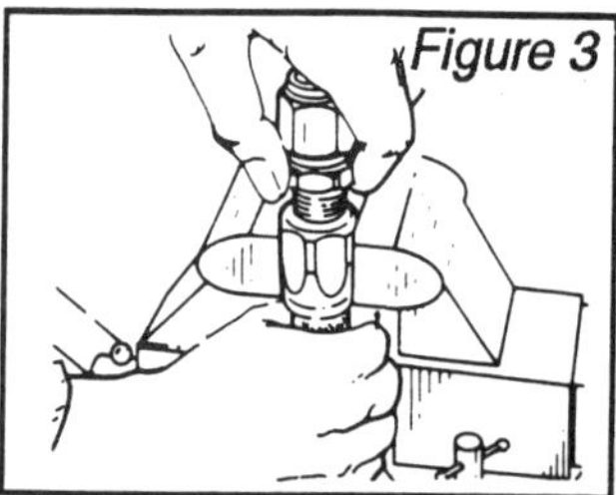
Figure 3

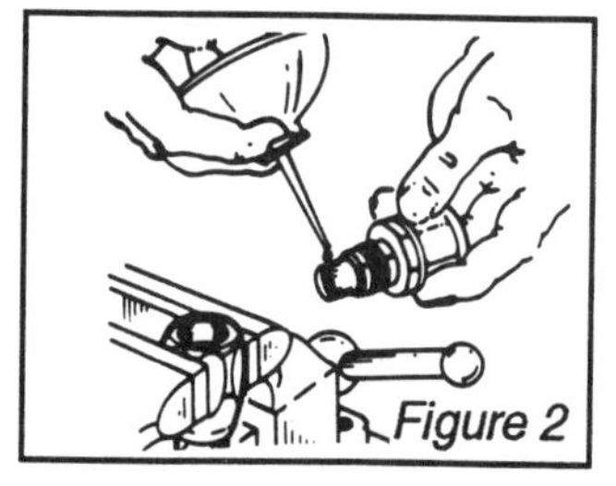
Figure 2

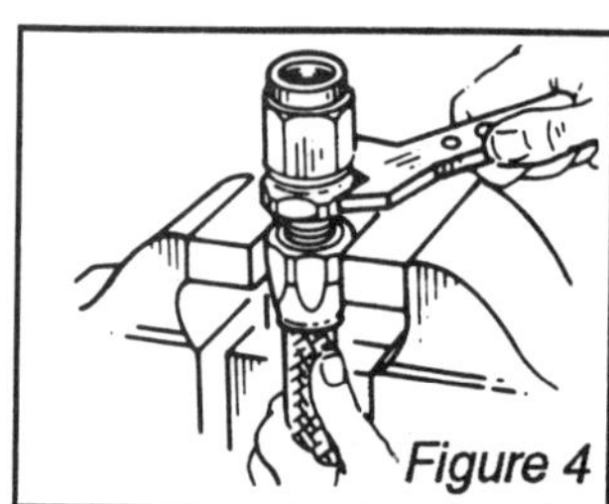
Figure 4

Drawings courtesy of Carroll Smith and Earl's Performance Products

Chapter
5

The Race Car Body

Designing A Race Car Body

Designing a race car body begins with the rules that a particular division or sanctioning body implies or enforces. Once you have determined what is legal and the type of body your car can run, you begin the design. Most race car body designs are based on OEM bodies like Chevy, Ford, Dodge, etc. So now that we know the type of body we will use, take what Detroit has already built and adapt it to the rules you will be using.

Most late model bodies use specific measurements and written instructions with diagrams instead of templates. (See drawings on next page.) This enables the race car body to have some design improvements built in because the body rules are not so strict without templates. You can build a body that conforms to measurements instead of templates and this allows you to design in aerodynamic advantages. One other area of design that goes into the body that we build is the removable panels for ease of maintenance and ease of repairability.

To meet these goals, you need to design a body that has simplicity, body panel replacement that's not difficult, panels that can be made from the smallest amount of material to help reduce cost, and a design that conforms to your sanctioning body's rules.

Aerodynamic Design

Aerodynamics is a major concern in body building because the more aerodynamic the body is, the faster the car will be. Not only is the car faster but the air is free and any advantage you can pick up here doesn't cost you a thing. So use the air to your advantage. Use each and every rule to its maximum.

Use smooth rounded edges so it flows through the air with ease, making sure that you gain maximum downforce for traction with what you have avail-

Notice how this body is styled to provide a more aerodynamic design. Edges are smooth and round, and the roof is shaped to direct air to the rear spoiler. Skirts are used to keep air from flowing underneath the body.

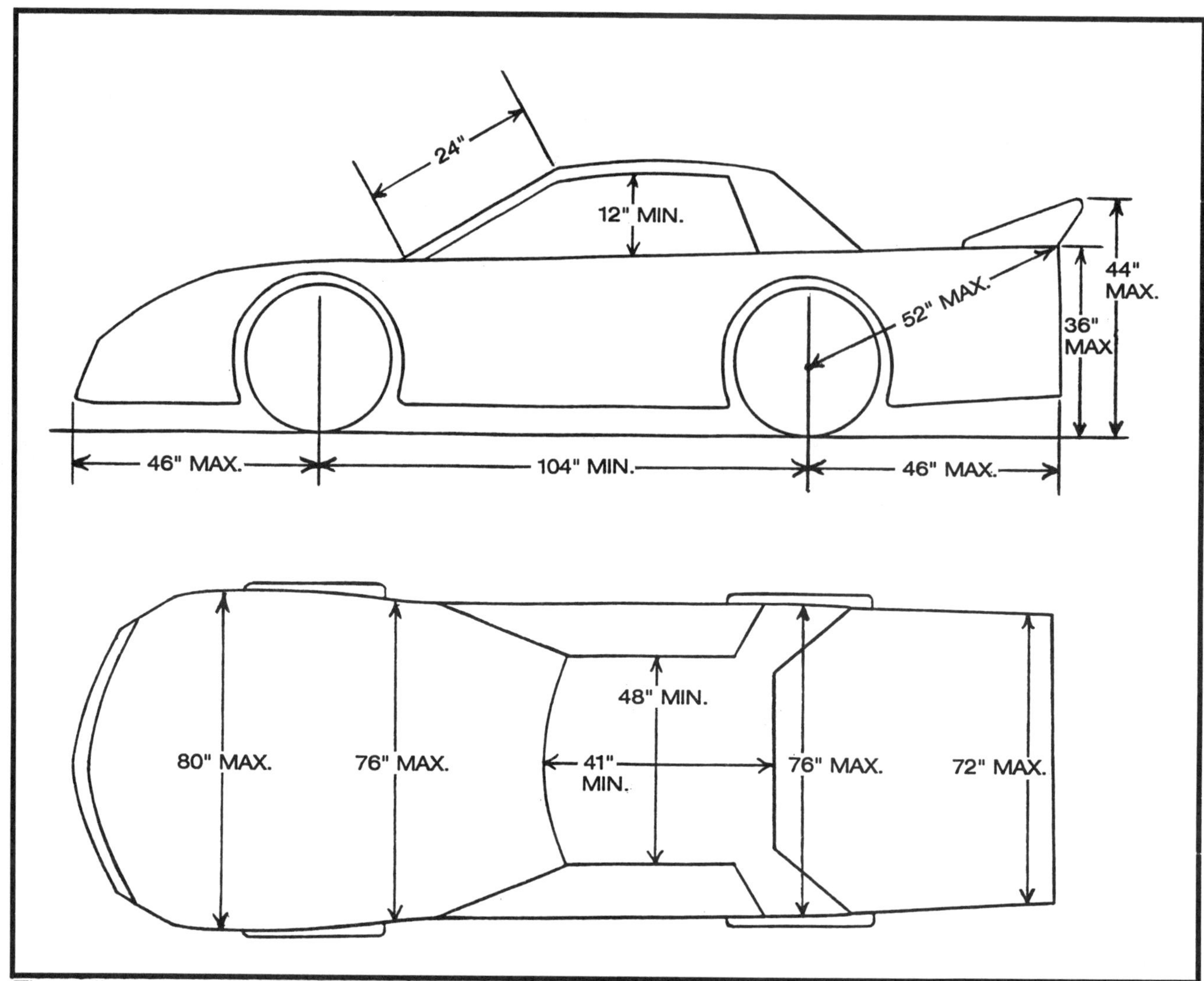

These dimensions are typical of those required for many late model style cars. These do not represent any particular racing association rules, so check with your track's rule book for details.

able. Late model dirt cars need all the air they can get directed toward the rear spoiler for traction at the rear wheels. The roof line can be used to achieve this. Since dirt cars don't have windows, the air that goes over and around the roof needs to be directed to the spoiler.

Try to keep as much air from going under the car as possible. Use skirts and mount body panels as low as the rules allow. One of the best ways to help you incorporate aerodynamic designs into your car is to look at what the Winston Cup cars and Trans Am cars do. You can learn a lot by studying pictures or looking at these type of cars when you have a chance.

Another way to help you determine aerodynamic design is to watch your car on a hazy or foggy night or when the dust is more than usual. You can usually see the path of the air flow over and around the car. You can use this knowledge to help make a more aerodynamic car. I have seen some teams do some testing using yarn taped all over the body and interior to monitor air flow visually. Any of this will help. Aerodynamic styling will aid a dirt stock car on short tracks, so anything that you can do will help improve your car's lap times. If you don't believe it is important, just put your hand out the window of your car going 60 MPH and try to hold it flat, versus turning it sideways to flow with the wind. If that little area puts that much force on your hand, imagine what an 8" x 72" spoiler will do for a race car.

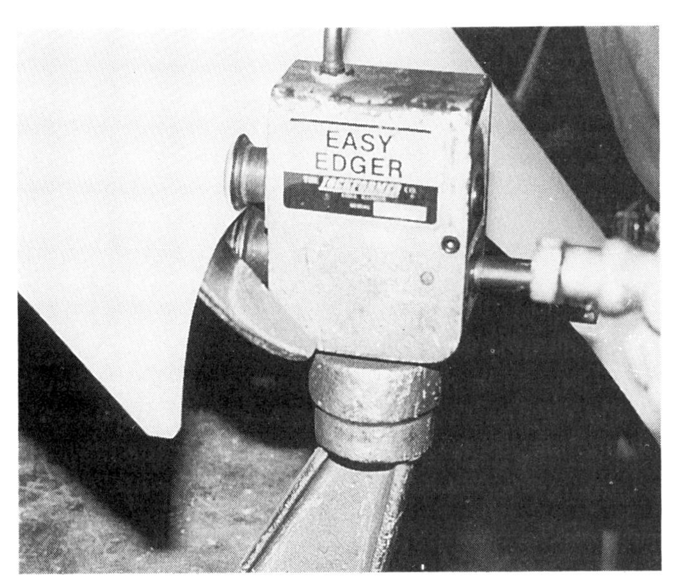

Using an Easy Edger to create curved or radiused edges.

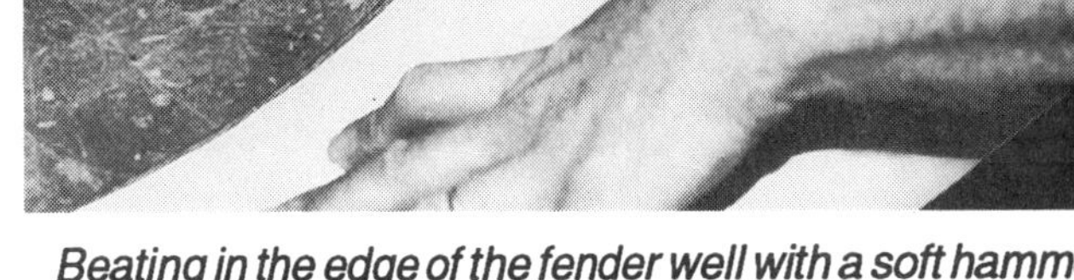

Beating in the edge of the fender well with a soft hammer after Easy Edging. This creates a nice finished edge.

Body Materials

Materials that are used in building late model bodies consist of .063-inch, .040-inch, and .032-inch thick aluminum sheets that are prepainted and very workable. This material is available from race car manufacturers, aluminum suppliers, sign companies, or the manufacturer — Wrisco Industries Inc. The .063-inch is used for floor pans, firewalls, and driveshaft tunnels. The .040-inch and .032-inch are for body parts and interiors. 1/8-inch and 3/32-inch thick Lexan is used for spoilers and dashboard side deflector panels. .090-inch and .110-inch plastic is used for skirts and nose panels. This plastic is readily available and comes in a variety of colors to match the aluminum that is pre-painted. Most race car parts houses or manufacturers carry this material.

You will also need a variety of rivets and washers to assemble bodies. You will use 62, 64, and 66 aluminum rivets. You will also need a variety of aluminum straps and angles. The most common used are 1/8-inch to 1/4-inch thick aluminum strap that is 1-inch to 1.5 inches wide. The aligning angle material used will be 1" x 1" x 1/8". 3/4-inch aluminum hinge is used to mount spoilers for easy angle adjustment. An array of Dzus fasteners and small 1/4-inch bolts are also needed for assembly.

Fabricating The Body

Fabricating the body starts with the sheets of aluminum that are first cut to the basic size of the panel to be used. Using patterns to lay out body panels on the sheet of aluminum you are using will let you determine the most efficient way to use all material without waste. After you have done this, start cutting panels with a stomp shear or hand shear to their basic size. After you have the basic piece sheared to size, cut wheel well notches for the body mount overlap.

The first thing that is done after each panel is cut out and trimmed to exact size is to use an Easy Edger to create any curved or radiused edges like wheel wells. What this means is that you take the metal and make a rough edge into a doubled edge that is smooth and strong. This process is done first by running the metal through the machine which produces a 90-degree bend 1/4-inch wide. Then flatten that bend with a wooden mallet or body hammer.

You must be careful not to ding the area of the panel that is away from the existing break. In other words, watch for hammer tracks.

Once this procedure is done you are ready to perform the breaks that form the edges of each panel. On each body panel you will break over the top and bottom edge and then flatten them out so the edges of the panels don't have the single cut edge. (This is the same procedure as the Easy Edger, but performed with a sheetmetal break on straight edges.)

A bench-mounted crimper makes a more precise and accurate bend with more durability. It also serves as a shrinker/stretcher.

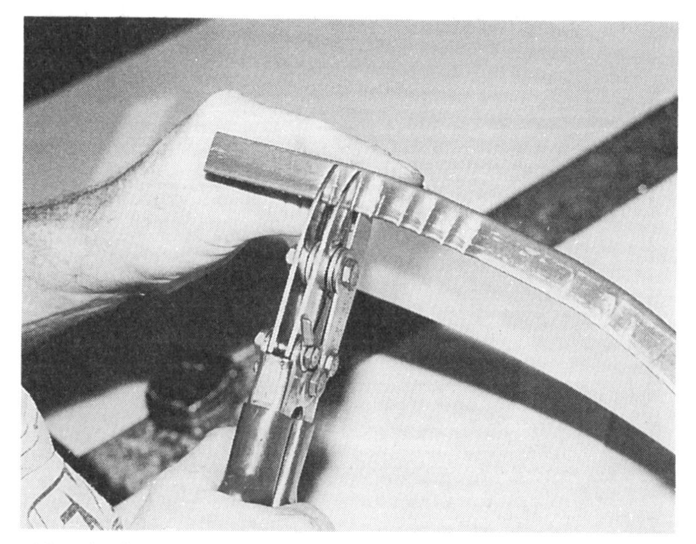

Hand crimpers are used to create a bow.

One last step to form body sides is to make a series of breaks 1/8-inch to 1/4-inch apart at the top that gives the same effect as rolling the body over a rounded edge. This edge is used to overlap the interior.

Anywhere there is a rough edge, there needs to be a hem or double edge. A hem or double edge is created by folding over the edge of the aluminum to create a smooth, rounded edge on it instead of a sharp, rough edge. Do this by using a sheet metal break to create a quarter break, then smoothly flatten the metal out.

Now you can either put a seam or a break down the middle of the body or just pound it (put a bow in the middle) to give the car a stock appearing look.

These are the basic fabrication techniques included in building a body. Interior and deck panels are fabricated in a similar fashion but use edge breaks that are 90 degrees. They are put up against each other to help support the interior panels and keep them from sagging. Usually interior panels are made anywhere from 16 to 24 inches wide and about 60 inches long. When the interior panels are built, be sure to check for squareness and fit before you make any breaks in the material. Double checking yourself in this way will help reduce a lot of remakes and wasted material.

As we touched on earlier in the book, fabricating all body panels and interior parts becomes much easier once patterns have been developed. One other thing is that, as in any other job, having the proper tools at hand allows fabricating to be a simple job. To make the fabricating job easier, you should have the following tools to help you:

A finished roof panel. The shape of the outside pieces create the bowing.

1. Hand shears – left and right
2. Long hand shears (scissor type)
3. Electric shears
4. Hand crimper (to form metal that has a break on the end)
5. Shrinkers
6. Hand pop rivet gun

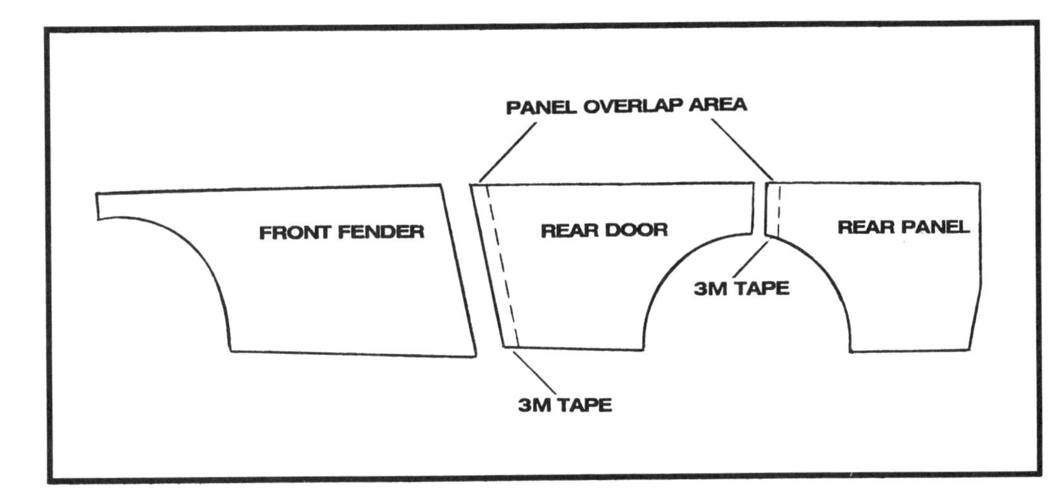

3M VHB tape is used to bond the 1-inch overlap areas of the body together. Run the double-sided tape the entire length and width of the overlap area.

7. Air rivet gun
8. Easy Edger (puts edge on a radius)
9. Good quality sheet metal break 6-foot to 8-foot wide
10. Stomp shear
11. Bead roller (makes strengthening beads and improves panel appearance)
12. Hand drill
13. A variety of drill bits including vari bits and hole saws
14. Clecos
15. A variety of vise grip clamps

Sheet metal work is a tedious job and you must learn to take your time and do it right. I've seen some rough sheet metal work in my days and this is one area that really brings out the craftsmanship and pride in one's work. One of the best ways to learn sheet metal work is to pay attention to what is going on around you. Go to a race track and look at the different styles and quality of work. Then apply yourself and make your job the best one yet.

Hanging The Body

When hanging the body you need to first assemble the entire side on a table. Attach the front fender panel section to the rear door section. Use the 3M VHB tape to bond the 1-inch overlap areas together. Run the double-sided tape the entire length and width of the overlap area and press together. This leaves a rivet-free attaching part which simplifies assembly and decal work. Once you have the primary pieces put together, start assembling the skirting that is put along the bottom of the doors. When installing the skirting we use the 1" x 1" x 1/8" aluminum angle as the backing for the skirt and this serves as the attachment point for the straps that support the body at the bottom and middle.

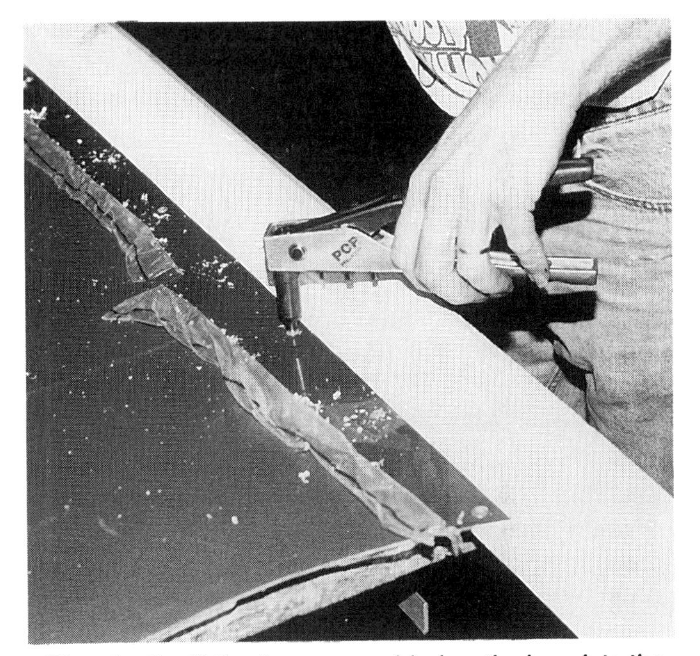

The plastic skirting is preassembled on the bench to the bottom of the body side panel.

Get all of your body panels assembled with all the skirting and support angles in place. This makes the next job of installing the body much easier because when you hang the body on the car, all the panels are already assembled. This enables you to mount the body at the front, middle, and back with Clecos. Now you can string the car's body line or eye it to make sure that the body is true and straight.

While assembling, remember that the measurements contained in the rule book must be met. So check your measurements before you start drilling

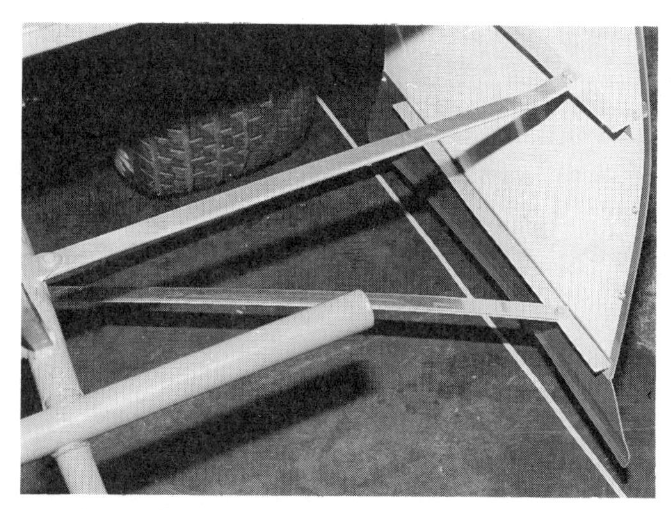

Use 1" x 1" x 1/8" aluminum angle as the backing for the skirt and the body panels. This serves as the attachment point for the straps that support the body at the bottom and middle.

A recessed hood pin plate mounted with double-sided 3M VHB tape.

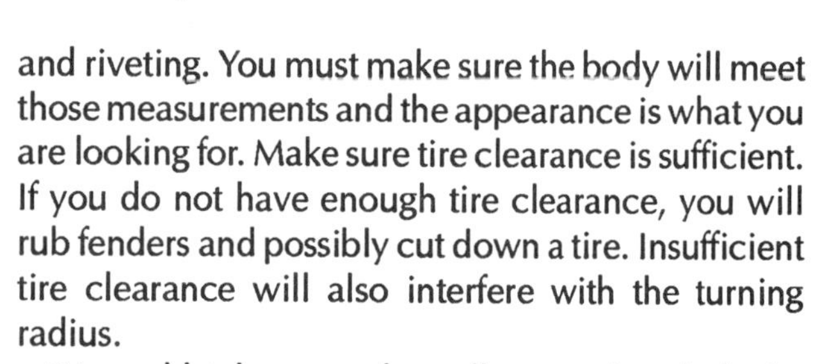

and riveting. You must make sure the body will meet those measurements and the appearance is what you are looking for. Make sure tire clearance is sufficient. If you do not have enough tire clearance, you will rub fenders and possibly cut down a tire. Insufficient tire clearance will also interfere with the turning radius.

We weld tabs onto the roll cage, then bolt the aluminum angle to the tabs. Then the body is riveted to the angle. This keeps from drilling a lot of holes in the roll cage tubing and prevents water from getting into the cage.

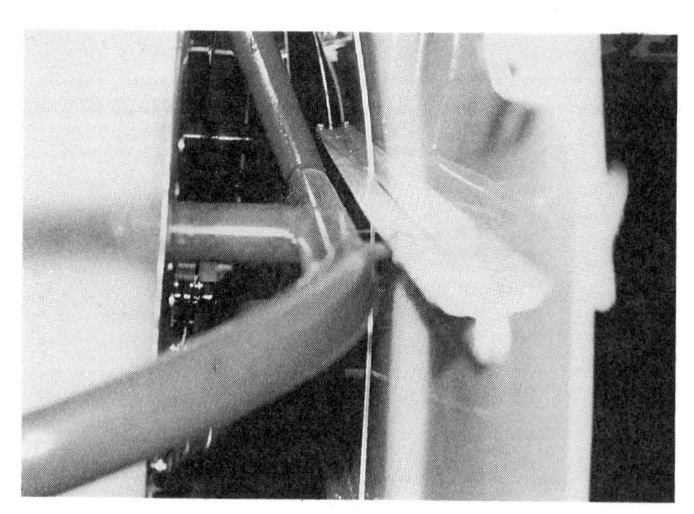

Weld tabs onto the roll cage, then bolt the aluminum angle to the tabs. Then the body is riveted to the angle.

A trick that helps the roof stay on the car is to apply silicone along the front and rear hoops of the roll cage tubing where the roof sets against them.

Roof mounting is done just like the body mounting. Make sure that the roof is square with the body and is centered over the cage properly. One trick that helps the roof stay on the car and doesn't let the roof move around is to apply silicone along the front and rear hoops of the roll cage tubing where the roof sets against them. Then use rivets or bolts to hold down the roof with the roof tabs welded to the cage. Roof side panels and front posts need to be securely mounted to the roof edges and at the point where the sanctioning body's diagram shows.

Hoods are mounted with four hood pins. First cut the hood piece out, leaving extra material to double-edge it to create smooth edges. Then locate the hood in its exact location with the four hood pins already

Spoilers are mounted at the rear deck with a piano hinge so the angle can be changed for downforce adjustments.

An air-powered rivet gun pulls the rivet a lot tighter than if a hand-held riveter is used. This helps durability.

mounted in place. Carefully tap the hood on top of each hood pin. This will leave an indentation in the metal to show you where to drill the holes for the hood pin openings. After you drill the holes, set the hood back on the car and locate the hood scuff plates and rivet those onto the hood.

Cutting the hole in the hood for the air cleaner is done in the same manner as locating the hood pins. Use the carb air cleaner stud or float bowl vents to locate the center of where the hole needs to be cut out. Scribe a hole out with a string and a center punch, then cut out the air cleaner hole.

Nose panels need to be mounted on a pre-designed nose bumper that is made to fit the particular year model being used. Again the nose panel has measurements to go by and it should be centered in the car left to right. Most sanctioning bodies require at least 4 inches of ground clearance, and the nose can't be mounted further than 45 inches from the center line of the front axle to the tip of the nose. The nose filler panel (the part that is between the nose and the hood) should follow the contour of the nose running smoothly into the hood. The hood will usually contour into the overall body height at or near the center of the front wheels.

Spoilers should be mounted at the rear deck panel and be adjustable with a hinge so you can change angles for downforce adjustments according to track conditions. Forty-five degrees is a general starting angle for a spoiler. If the track is extremely dry and short, use more angle. If the track is long and tight use less angle.

When mounting bodies, make sure rivet spacing is about 5 to 6 inches on the body mounting, and 3 to 4 inches on interior panels. Always use backing washers and deburr the back side of the drilled hole so your washers fit snugly. Use 1/4-inch bolts and Nylock nuts to mount all aluminum bracing to the body. Make all body lines smooth and don't leave any jagged edges.

Chapter

6

Painting the Chassis

Painting the chassis is a difficult job and must be done properly if you want to have a professional looking result. There are several different ways to prep and paint a chassis. First, we will discuss the way we paint cars, and then talk about some of the other options.

Conventional Spray Painting

For years we used to paint with a conventional spray gun. This system works well, but we did not know what we were missing. Let me explain. The conventional spray system uses at least 1.5 gallons of paint and that is not including thinner and hardener. It would take about 3 to 3.5 hours to paint the car after it was completely prepped and put on a rotisserie. (The rotisserie is a system connected to the front bumper and rear bumper that lets the car rotate upside down — see photo).

Conventional spray painting takes a lot of patience because a painted chassis sometimes develops runs. You must learn the technique for painting chassis and roll cages. It takes a lot of practice to do a quality job. If you are going to learn to paint, practice on some roll cage tubing on an old frame to get the feel for it.

When painting, the most important thing is to give the overall chassis a good base coat. Start at the rear and work your way up to the front. A good paint job starts with the base coat. This is the most important step.

Once you have a good base coat, you can spray a little more paint on the second time around to help promote gloss. Start spraying the second coat on the bottom first, and finish by spraying the car right side up to cover all the top bars evenly. The reason is that if you cover the top last, all of your dry spots and mistakes will be on the bottom. The bottom of the car is the least visible and you can't see the flaws as easily. If you have some dry spots and you are on your last coat, you can go over the car with a 75 percent to 80 percent thinner/paint mixture and blend in the spots.

Notice the full safety equipment this painter is wearing. Painting is considered a hazardous activity. Use proper safety equipment.

Using the correct thinner and hardener makes a big difference in properly finishing a chassis. We use Ditzler PPG acrylic enamel and use #601 or #602 thinner, depending on the ambient air temperature. If it is cold, use a faster drying thinner. If it is hot, use a slower drying thinner. Always use a hardener for the best gloss and durability.

Using electrostatic spray painting, the paint is attracted to the metal and is pulled around the roll bars.

The chassis is mounted in a rotisserie for ease of painting both sides. Note the ground strap connected to the chassis for the electrostatic system.

Electrostatic Spray Painting

Now that we have discussed the standard painting system, I will share with you the best thing that ever happened to painting a chassis. This system probably would not be practical for occasional painting but if you are going to paint often, make the investment. It is called Electrostatic painting.

This system cuts the time and material in half. What was taking three hours now takes one and a half hours. What was taking one and a half gallons of paint now takes 3/4 of a gallon. It is simply amazing! You basically paint with the same process, but now the paint is being electrically attracted to the metal and is being pulled around the roll bars. You are actually painting the backside of the tubing while you are painting the front. The paint wraps around the tubes. In essence, once you have the bottom coat on, when you go to the top you already have the top primary coat on. This is all done by charging the paint. The chassis is grounded so when the paint leaves the gun, it pulls it to the metal.

As you can see, speed and efficiency has drastically been increased with this system. This type of painting also takes some getting used to because the spray patterns change from that of a normal spray gun. If you get the opportunity to use one, practice with it on some old pieces of tubing or an old chassis.

The Prep Work

The most important part of the paint job is how you prep the metal surfaces that are going to be painted. Chassis prep involves cleaning, deburring and grinding the chassis so the paint will adhere to the metal and the finish will be smooth.

After a chassis is welded, you will notice that there is splatter around some of the welds and in tight areas where the welder tip barely reached. There may also be some welds that don't overlap well and might

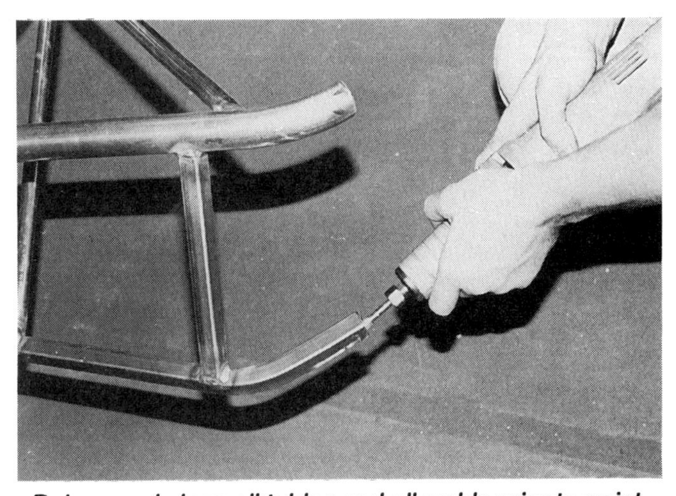

Deburr and clean all tubing and all welds prior to painting.

produce a drip type weld. All of these must be deburred and ground down smoothly. Look at any surface that sheet metal will lay against to make sure welds are ground down smoothly. Motor mounts need to be cleaned to provide a smooth surface for the motor to set against.

Deburr and clean any tubing that has been left open on the ends like body mounts, front bumper ends and right side door bars. Make sure that all edges and corners are rounded and smoothed. You don't want your chassis to look like it has been painted over a piece of sandpaper with rough edges.

The next step is to wire brush any tight areas where two bars meet. This cleans up the weld itself and loosens any residue that has been built up during welding.

After you have completely gone over the chassis with this process, you are ready for the wipe down stage. When you are wiping down the chassis, it is a good time to check for any missed welds or braces, gussets, etc. We use clean up thinner to thoroughly wipe down the chassis from front to back, keeping rags clean and changing to clean ones every so often. This removes all the grease and oil film from the tubing and creates a good clean surface ready for paint. Be sure to keep thinner-soaked rags picked up and properly stored away from sparks and flames.

Priming

Now you asked about priming? We have tried just about every type of metal primer possible. The end result is that it doesn't make any difference on the finish and durability of a dirt car chassis. Believe me we have tried many primers, and after one or two races, the result is the same. The chassis becomes pitted and looks like it has been run through a bead blaster. The primered surface would not have done any good. It is not worth using a primer on the chassis because of the abuse it takes.

Chassis Color Choice

The acrylic enamel paint sprayed on with hardener and applied to a good clean surface has a very glossy and durable finish. Since a chassis takes so much abuse, there are a couple of things to remember about colors. Use a color that touches up easy because you will be making repairs that require repainting. And, the general wear and abuse from dirt requires occasional paint touch up. Usually any lighter color makes cracks and breaks more visible for repair. It does not matter what color you use, but darker colors tend to hide or camouflage cracks or breaks. A darker colored chassis requires more attention during visual inspection to locate these problems.

Always take the time to touch up the paint after repairs. This keeps the chassis much more presentable and new looking. Paint bolt-on bumpers, door bars and any other items that you want the same color as the chassis while you are painting the main frame.

Powder Coating

Powder coating is another type of painting process that racers sometimes use for the chassis. This is a process where electrostatically charged colored powder is sprayed on the grounded chassis. The paint is then baked on in an oven. It produces a paint job that is definitely top quality. There are no runs or dull spots, and you get a very durable paint job. There are not as many colors available and the expense is greater than conventional painting, but this is definitely a good option for a racer who is extremely particular about detail.

Powder coating does have some drawbacks, however. When repairing a damaged car, the area to be welded must be cleaned well. The welder does not like powder coating and it produces a terrible odor when welded. Repairs are a little more difficult and inconvenient. Once a repair is made on the chassis, the area has to be powder coated again. Touch up sometimes is not a good match to the original application.

We do not have anything against powder coating, but you have to be prepared to pay more for the job. And you have to transport your chassis to and from the place that performs the job. Powder coating can prove to be an extra cost in terms of time and inconvenience as well.

One other item to consider is that powder coating will help normalize the metal. This is because the baking process temperature reaches 200 degrees or better. This would be good for 4130 tubing.

Chapter

7

Front Suspension & Steering

Designing A Front Suspension

Designing a front suspension and laying out what you want is quite a project to take on unless you are a design engineer or know someone that is. We suggest you use a current, already proven design. If you choose to design your own, base it on one that is being utilized today, then try to make improvements from there.

What is used in most dirt late model front ends is a design from the Ford front clip with lower strut rods going forward. If you have access to a computer and use a program designed to determine roll centers and front end geometry, then you will notice that when you start moving things around it changes something else. Many of the design elements are dependant on one another. So if you choose to try to improve or redesign a front end layout, start with the basics. Start with a design that is currently winning and find ways to improve on that design.

Component Choice

When choosing components for the front suspension, you need to determine the weight, durability and safety of all of the components you use. First, remember that anything which revolves or rotates should be as light as possible, but there must be reliability and a safety margin built in. When you go past the point of reliability and safety, then it is too light.

For instance, you don't want to build a component such as a spindle that is so light that the steering arm bends or breaks. If it does, you can't finish the race, or you hit the wall or somebody else. A critical part that is too light is dangerous and not practical. The rule of thumb is always the same — any rotating and unsprung weight need to be as light as possible. And, rotating weight is the most critical.

We have found that using a magnesium hub rather than an aluminum hub saves 2.5 pounds per wheel. Then combine that with a drilled 1.25" x 11.75" rotor and you save another 2.5 pounds per wheel. That is five pounds per wheel times four wheels, which equals 20 total pounds of rotating weight saved. The mag hubs and drilled rotors are proven and reliable,

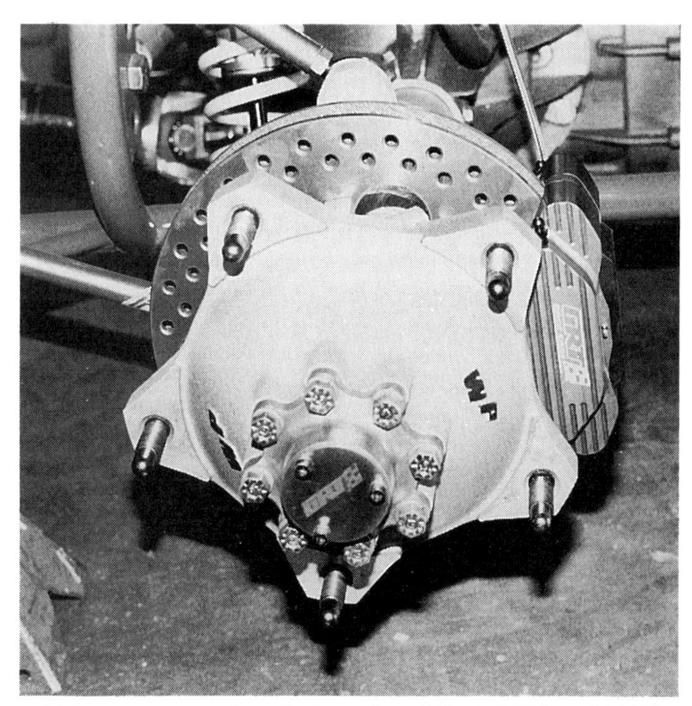

Using a magnesium hub and a drilled 1.25" x 11.75" rotor saves 5 pounds per wheel.

Mounting a shock upside down transfers the heavier part of the shock weight to the chassis, which is sprung weight.

so this is a must. Do not sacrifice the extra weight here!

Other ways to save on suspension component weight is by using aluminum rod end bearings, aluminum trailing link tubes, aluminum wheel hub nuts, and aluminum shocks that are mounted upside down. They are mounted upside down because the lightest end is the shaft end and it will be the one that is moving. Mounting it upside down transfers the heavier part of the shock weight to the chassis, which is sprung weight. AFCO offers aluminum shaft shocks and carbon fiber rod ends for real weight savings.

To save additional weight, we also use drilled wheel studs and titanium lug nuts. Spindles are made of lightweight steel and are gusseted well for strength. The spindle is one area where you don't want to get too light because of performance and safety.

Use the suspension components that your chassis builder recommends because they know what works and what is safe. There are other areas where you can save weight, but for the cost versus what is practical, these are all that we suggest you use. In other words, spending lots of extra bucks to get the components any lighter is not feasible for the type of racing we do. We will talk about one other major expensive component in rotating weight – carbon rotors – in the brake chapter.

Choosing A Spindle – Design Criteria

When choosing a spindle design you must look at what performs best in several areas. The steering axis inclination angle affects static camber and loading and unloading of the wheel during cornering, along with caster setting. The steering arm length and location determines Ackerman steer, steering quickness and bump steer. The spindle height helps determine roll centers due to the angle of the A-arms which are connected to the spindle at the top and bottom.

The steering axis is the line drawn through the upper and lower ball joint centers about which the spindle rotates as it is steered. The inclination angle of the steering axis is the angle in degrees between the steering axis and true vertical. This angle has effects on camber, steering and how much a front end loads and unloads during cornering. I've seen everything from 5 degrees to 10 degrees used on a dirt car. What we feel works best on our cars is 7 degrees. The more inclination angle used in the right

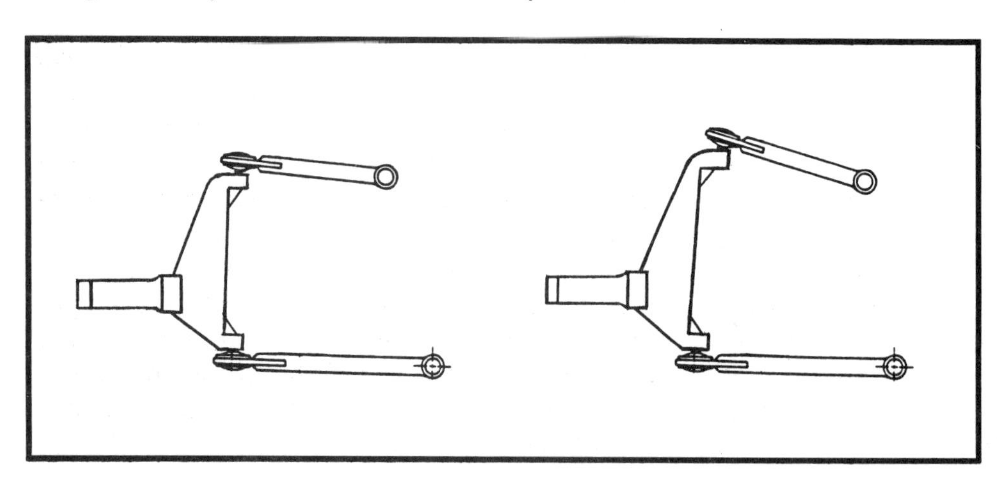

The spindle height helps determine roll centers due to the angle of the A-arms. The steeper angle of the upper A-arm (right) creates a higher front roll center, with all other factors being the same. The steeper upper A-arm angle also creates a faster camber change curve (more negative camber gain per inch of bump travel).

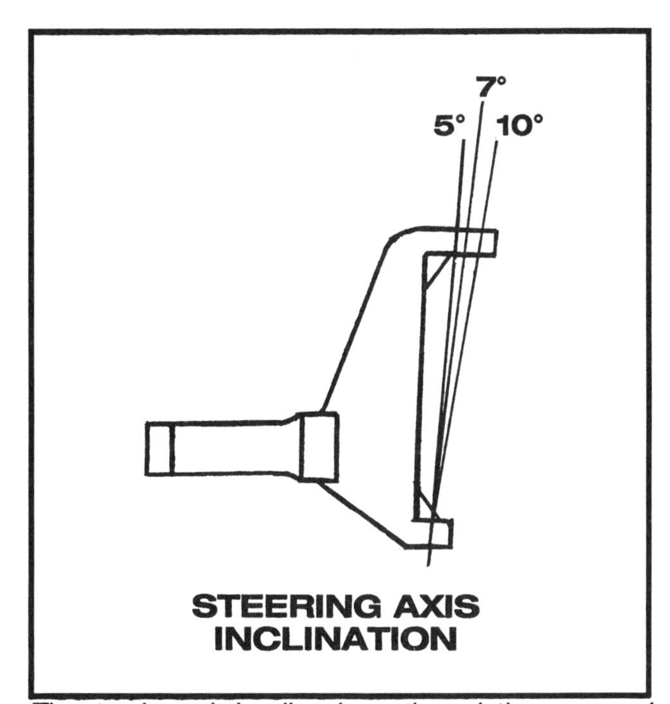

The steering axis is a line drawn through the upper and lower ball joint centers about which the spindle rotates as it is steered.

front spindle, the more it will load the left rear when car is in a counter steer situation during cornering (front wheels turned to the right).

The caster setting produces this loading effect, but the more spindle axis inclination present, the more it multiples the rate of loading the diagonally opposite corner. For an example of what I am talking about, scale your car with the front wheels straight ahead and take a weight reading. Then turn the front wheels to the right 20 degrees and take another weight reading. The left rear and right front will have gained weight. If you want to do this comparison between a 7-degree and 10-degree spindle, you will find that the 10-degree spindle will produce a greater weight gain at the left rear and right front when the wheels are turned.

I have had drivers ask me to build spindles with different inclination angles for different types of tracks or set-ups. But we feel the 7-degree spindle inclination produces the best performance all around. Too much inclination angle makes steering more difficult, while too little angle makes the steering feel easier and have a darty feeling. The less inclination angle there is, the less caster effect the spindle has. And, caster creates front end directional stability, so less inclination angle makes the steering feel lighter and have less steering stability.

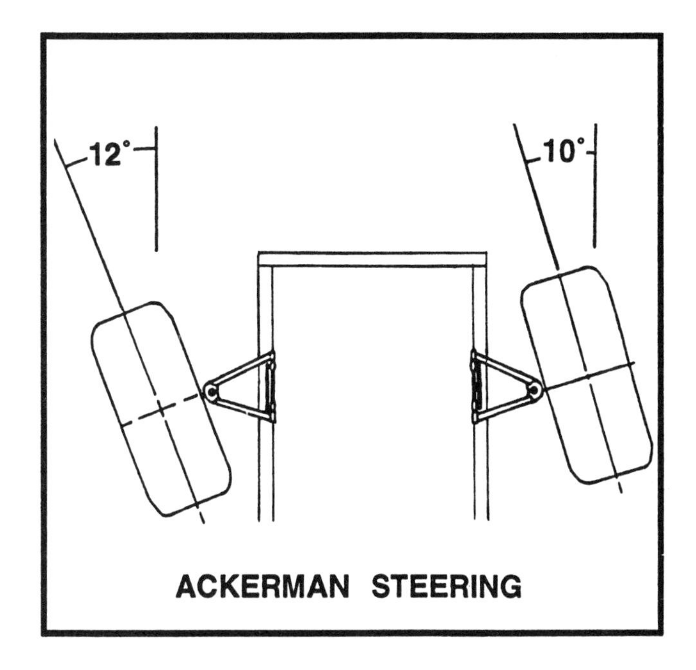

Steering arms help determine the amount of Ackerman steer and bump steer in the front suspension. Ackerman steer is a steering geometry designed to make the left front wheel steer at a greater angle than the right front when the car is steered to the left. In my opinion, Ackerman is not critical to have on dirt late models and we feel that about one degree of Ackerman is sufficient. This makes the inside wheel turn one degree more than the outside front wheel. The closer the steering arms are to being parallel to the center line of the car, the closer the Ackerman will be to zero. You can check your Ackerman with a good set of turn tables. Set the wheels straight ahead and then turn them 20 degrees to the right. Check the angle of both sides. One wheel will turn more than the other. This is the Ackerman steering angle. Try to have the Ackerman as close to zero as you can.

Bump steer is affected by the steering arm's vertical attachment point to the tie rod. Bump steer is adjusted by placing shims between the bottom of the steering arm and the tie rod. If the steering arm is properly designed, the bump steer shimming required will be minimal or none at all.

The steering arm length affects the steering quickness. The shorter the arm is, the faster the steering is. The longer the arm is, the slower the steering is. We have found that too short of a steering arm makes the car more jerky. It is harder to be smooth and this is quite noticeable on longer tracks like Eldora,

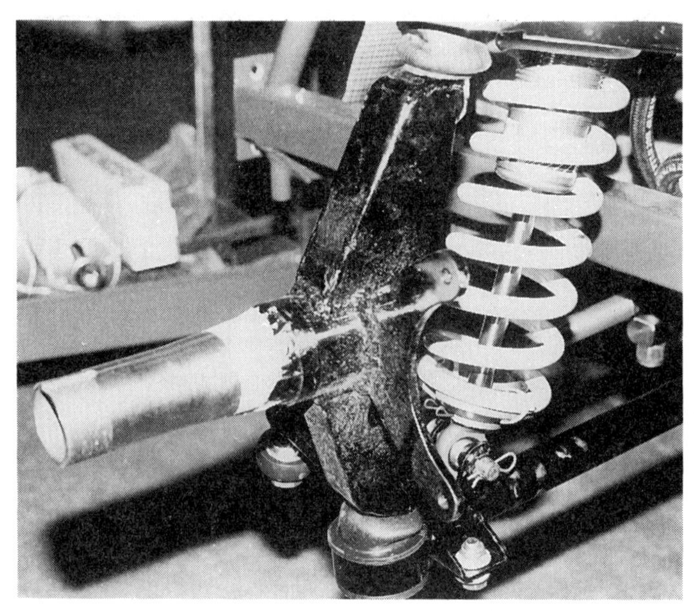

When comparing cost, reliability and repairability, lightweight steel is better than aluminum as a spindle material.

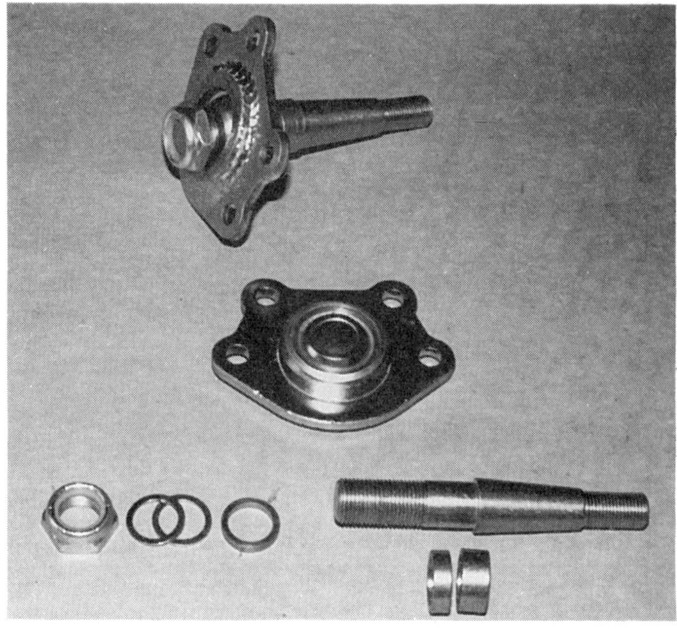

A variable length pin upper ball joint, such as this one from Coleman Machine, can be used to adjust the roll center by moving the upper pivot point up or down.

Pittsburgh, etc. We have found that a steering arm in the 5 to 5.5-inch length works best and definitely makes the car smoother.

Spindle height is the last area of concern. The height of the spindle determines the mounting angle of the upper A-arms. This greatly affects roll center location in both height above ground and left-to-right location. The height of a spindle must be taken into consideration with the layout of the entire front end design. You can raise and lower roll centers with spindle heights and move it left to right. But even though you get a roll center where you think you want it statically by adjusting spindle heights, you still have to work with the entire front suspension design to ensure that the roll center and camber change curve work properly during body roll. Changing spindle heights is a good way to work with roll center location without moving inner frame mounting points. Most spindle heights range from 8.75 to 10.25 tall.

In designing our spindle height, we took into consideration our target roll center height (5 inches above ground) and lateral location (1-inch to the right of vehicle centerline), plus upper and lower A-arm lengths and inner mounting positions. All of these items must be considered together to yield a proper roll center location, and a roll center that does not move around drastically while the body rolls during cornering.

Spindle Materials

Once you have determined the spindle design you want, then choose the proper material. We firmly believe the fabricated steel spindle is the best system. When comparing cost, reliability and repairability, lightweight steel is better than aluminum. Aluminum is costly, bulky and doesn't save much weight because of the way you have to build them to be durable. Stay with the fabricated steel.

Ball Joints

Ball joints are an area of front ends that sometimes are overlooked or neglected. We suggest you use a good quality ball joint such as Moog or TRW. The most frequently used upper ball joint is part number K-6024. The most common lower ball joint is either the K-727, which is a heavier screw-in Chrysler type, or the K-6141, which is a lighter press-in ball joint.

Adjustable monoball joints are another option for use as an upper ball joint. But this is another area of cost versus what is practical. They can be used to adjust the angle of the upper A-arm to change roll enters, but again they are more expensive. And, if your car is designed correctly, you don't need an adjustable ball joint. The main thing you need to remember is that whatever ball joint you use, keep

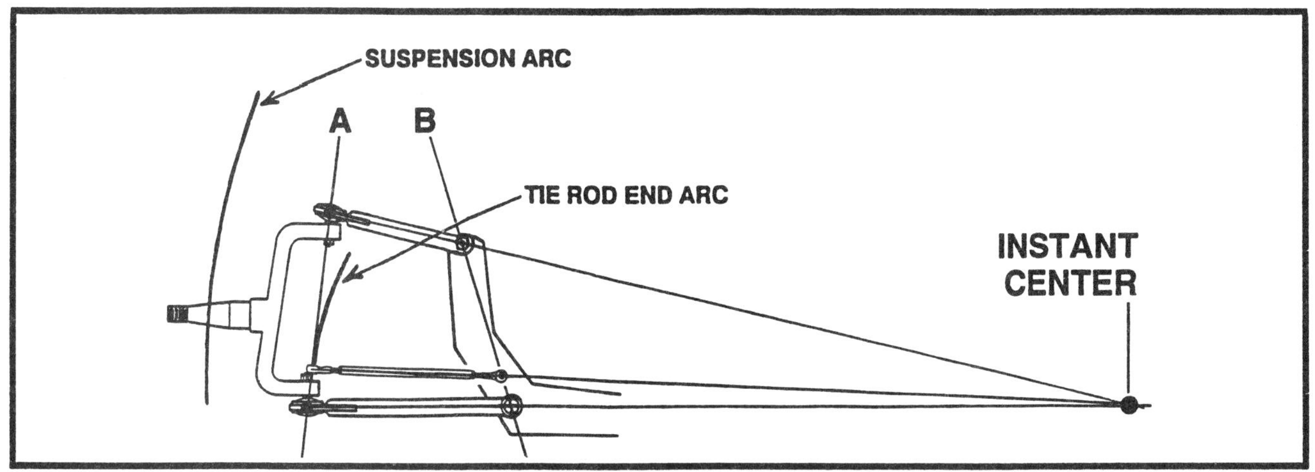

Bump steer is zero when the suspension arc and the tie rod end arc are parallel to each other. To accomplish this, the tie rod length must fall between the confines of planes A and B, and the tie rod end's centerline must intersect the instant center. The tie rod may be laterally displaced so long as the length remains the same as if it were between the A and B planes.

Bump steer can be corrected by shimming the steering rod end up or down on the steering rack or spindle.

a constant check on them for wear and proper lubrication. This is a critical handling and performance area.

Steering Rod Ends

We use aluminum 5/8" x 5/8" rod ends on our steering arms. They are light and strong. They make adjusting bump steer and toe-out easy. They don't bend like a regular automotive tie rod end, therefore they keep the front end more reliable and to specifications. Steering rod ends should be no smaller than 5/8-inch and a top quality brand. Keep them well lubricated (unless they have a self-lubricating liner) and check often for wear and slack.

Hubs

The wide five is the most common type of hub used on most dirt track cars. Any lightweight aluminum wide five hub from a quality manufacturer will give years of service provided they do not get damaged during a crash. The best way to gain performance is through light rotating weight. You can spend a few more dollars and get the wide five magnesium hub which is about 2 to 2.5 pounds lighter than aluminum. This is what we install on each late model we build. The Sierra mag hub is our choice.

Bump Steer

Bump steer is the change in steering angle of the front wheels caused by the front suspension moving up or down through its travel. Bump steer causes the introduction of toe (either in or out) into the front wheels when the suspension goes into bump or rebound. Bump steer occurs when the tie rod end follows a path different from the path the wheel is following. If you visualize the arc created by the front wheel as it moves up and down through its travel, along with the arc created by the tie rod end, you will see that both arcs must have the same instant center or else the tie rod end will move in or out in relation to the wheel, causing a change in toe. If the arc created by the tie rod end and the wheel are parallel, no toe change will be present. Bump steer results from the steering tie rod moving in a path which is dissimilar to the path in which the wheel it is connected to is moving.

Bump steer can be corrected on most cars by shimming the steering rod end up or down on the steering rack or spindle. A well designed race car will have no more than .015-inch of bump steer per inch through its suspension travel. We like as close as possible to zero at the left front, and .015 to .020-inch toe out per inch at the right front.

The Steering System

Passenger car steering systems are a lot more difficult to work with than the fabricated type. They are heavier and bulky. Adjustments can't be made nearly as easily because of the way components are bolted together, such as using a steering box instead of a rack. And with the passenger car system you have an idler arm and a Pitman arm plus a drag link. These are all heavy, wear out easily and develop slack. They are also hard to adjust for proper suspension geometry.

The passenger car steering is most commonly used in divisions like limited late models, super stock, or hobby stock. If you have to use a passenger car based design, then maintenance is the most important thing you can do. Tie rods bend easily and idler arms wear out. It is critical to check them each week.

One area of improvement in a passenger car steering system is the use of a steering quickener to speed up the steering ratio. They are most commonly available in 1.5 to 1 and 2 to 1 ratios. The 1.5 to 1 ratio improves the steering speed by 33 percent, while the 2 to 1 ratio improves it by 50 percent.

Steering Shaft & Universal Joints

The steering shaft and steering u-joints should be made lightweight and strong. The steering shaft should be designed in two sections (connected in the middle with a universal joint) for the safety of the driver. The shaft has to be designed so that it and/or the steering wheel is not pushed back into the driver in case of a heavy frontal impact. Building the shaft in two sections connected by a u-joint will cause it to fold up under impact. The shaft should be arranged so that when it is pushed from the front, the angled center section will push outward or downward. And, make sure that no u-joint in the steering system has an operating angle more than 30 degrees.

We strongly recommend using .75-inch OD D.O.M. tubing, .120-inch wall thickness for the steering shaft. Tubing is much lighter than solid stock. All of the major rod bearing manufacturers make an oversized .75-inch rod end (0.757-inch I.D. bore) for mounting steering columns made from .75-inch OD tubing.

The steering shaft should be designed in two sections (connected in the middle with a universal joint) for the safety of the driver.

The steering u-joints are available from several manufacturers. We use the Woodward brand. They make one for about every different spline found on a steering box, rack, or quickener. They also make smooth bore u-joints which can be used in mid-stream between the steering box or rack and the steering wheel in a multi-section steering shaft. These u-joints should be slid onto the .75-inch tubing about .5-inch to .625-inch. They will automatically be on straight because they are machined properly to align themselves. Then a good solid weld should be made all the way around the u-joint after it is trial fitted for length and location. Be sure to let the welded joint cool down at a normal rate.

Once the steering shaft is routed and installed, a quick release steering hub is installed on the cockpit end. These allow the steering wheel to be removed for easy driver entry and exit. This is especially important in a crash or fire to be able to get the driver out quickly.

Make sure you get a good quality quick release hub that either has a pin or 360 degree lock. The pin type is a little inconvenient in that you always must locate the hole to put the pin in and it is easy to loose the pin. We recommend the 360 type. And, always use a quick release hub made out of aluminum or steel — never plastic. Be sure to periodically check the parts for signs of corrosion.

Rack And Pinion Systems

When choosing a rack and pinion system, you must determine the ratio you need, whether power steering is going to be used, and the type of rack and pinion you want. There are several different brands available and most are very good systems. We have used most types available both with power steering and without power steering.

We feel that it is best to use power steering, so this helps us determine the type of rack that we use. When you use power steering you must have a servo valve (which regulates the power steering fluid delivery). This is another component that must be attached to either the chassis or the rack. The Appleton rack has the servo attached to the rack itself along with the slave cylinder, and this makes a very compact and functional rack and pinion system. There are less hoses to run and less u-joints involved routing the steering shaft. These are all factors we take into consideration when using the Appleton rack. The other rack and pinion manufacturers build fine systems also, but this is the most practical and efficient for our cars.

Another choice you must make is the servo stiffness which makes the steering effort easy or hard. Most are offered in a light (easy to turn), medium (moderate to turn), or heavy (hard to turn). Most drivers that run our cars prefer the light.

The steering ratio of the rack is also an important consideration. They range from 16 to 1 to 8 to 1. Most late model dirt cars use the quicker ratio and it is standard production on our cars. The 8 to 1 ratio is also referred to as a 3.4 ratio. That number refers to how far the rack eye moves with one turn of the steering wheel. The slower racks just don't seem to react quick enough for most drivers. We suggest you try different ratios and find what is most comfortable for you.

The mounting of a rack offers another advantage over other types of steering systems. The rack is easy to mount and most racks use the same three-hole triangular mounting pattern. They are mounted in front of the cross member centered in between the

The Appleton rack has the servo attached to the rack itself along with the slave cylinder, and this makes a very compact and functional system.

lower A-arm mounting points. The vertical location is usually mounted as low in the car as possible because you can always space the rack up with spacers if adjustments are necessary. The rack must be mounted with the center of the inner tie rods centered vertically with the lower inner a-arm mounting points. This needs to be relatively close but not absolute because you can perfect the location with shims under the rack or in between the steering rod end and rack shaft.

The rack mounting position in the chassis can also be changed to adjust for desired steering geometry changes. Moving the rack up or down affects the bump steer. Moving the rack back and forth (left and right) affects the Ackerman steering geometry.

Power Steering

As discussed in the rack and pinion section, the power steering systems requires a servo and a slave cylinder. These won't do any good without a pump to push the power steering fluid through the system. You need a good quality pump that is belt driven off of the crank pulley or mounted behind the bell housing driven off of a crank coupler. If you choose to use the standard power steering pump, the Sweet power steering pump is most common. They make a high and low pressure pump but we have found that the high pressure one works best on any system. These pumps are usually mounted on the front of the cylinder head with a bracket that allows easy adjustment for tightening belts.

IMCA modifieds usually use the block mount for the power steering pump. There are so many advantages to mounting the pumps to the rear that a pump manufactured by KSE is being used more and more. This pump can be mounted in-line with the oil pump and also includes a fuel pump. So now only one belt is required to drive the oil pump, power steering pump and fuel pump. This is a very practical system. You no longer have the pumps mounted on the front of the motor, and no fluid lines are running to the front of the motor. Motor changes have been simplified by this because you don't have to bolt and unbolt the pumps, and there is only one belt driving the whole system. (You still have the crank-to-water pump twin V-belt system, but you don't have to deal with those belts during a motor change.) Also, this moves all of the pump, line and fluid weight to the middle of the car instead of being out in the nose of the car. This system also uses a reverse-mounted starter so that is one other item which does not have to be removed and reinstalled during motor changes. As you can see there are many advantages to having the power steering pump mounted to the rear. It simplifies plumbing, motor installation, and weight percentages.

There are a few manufacturers (Tilton, Bert and Brinn) that make a good reverse-mount bell housing system that enables you to mount your pumps to the rear.

The power steering pump is simple to plumb because it has only two lines — the supply line and the pressure line. The supply line is usually a #10 in size and gets its fluid from a quart-sized power steering can that is mounted near the pump on a firewall or roll cage tube. A #6 line is used for the pressure side line. This line goes directly to the servo. Most servos have a "P" for pump to designate where the pressure line hooks in. The servo in turn has three more connections which include "T" for tank, which is the return line to the tank, and "L" for left and "R" for right. These line connections direct fluid to turn the car left or right when the steering wheel is turned in that direction.

The KSE reverse-mount oil/power steering/ fuel pump drives all off of one pump.

Remember when plumbing the system to use hose which is specifically designed for the high pressure and heat of racing power steering systems. We use the Earl's Power Steering hose, which is a triple steel and fabric reinforced high pressure hose. We use a #6 size for pressure lines, and a #10 for supply lines.

Power steering fluid is often overlooked. Always use a true power steering fluid in the system and not transmission fluid. The best fluid we have found is Royal Purple EZ steering fluid. It is designed for the extreme heat and has good anti-foaming qualities. Talk to your power steering manufacturer about their fluid recommendations.

Always keep an eye on the power steering fluid level. Change the power steering fluid often. Any time it has a burnt smell or appears black in color, change it.

Excessive fluid temperature is the leading cause of power steering component failure. Power steering coolers and filters are a good option to consider in any power steering system. Appleton makes a good filtered cooler can and AFCO makes a good in-line power steering fluid cooler.

Chapter
8

Rear Suspension Systems

There are a variety of rear suspensions being used today in dirt stock car racing. We will discuss all of them, highlighting their advantages and disadvantages. Through the 14 years we've been involved in racing, we have seen about every conceivable design, and have tried most at one time or another. Our primary success has been with the 4-bar, but we will try not to be partial when discussing other types of suspension. All of these different types of rear suspension systems have won races somewhere at sometime, so they all have a legitimate purpose. However, some types will deliver better results than others.

The rear suspension system is the most critical element on a dirt track race car. It determines how a car corners and how it is propelled forward. It is, in essence, the controlling element of the entire race car.

Leaf Springs

This type of suspension is as basic as they come. The springs are mounted so they perform three jobs at once. They serve as the trailing link of the car, providing the forward momentum hookup between the tires and the chassis. They also locate the rear end laterally much in the same manner as a Panhard bar. A third duty they perform is to provide the actual spring rate of the suspension. If it is a multi-leaf, they provide all of the spring rate. In the case of a monoleaf, they add partially to the total spring rate.

Most late model cars that run leafs are of the monoleaf (one single leaf) design assisted by a coil-over assembly. Usually the monoleaf spring rate is about 25 pounds per inch. Multi-leafs are more common on sportsman cars or street stock cars. Usually their spring rates are rated according to the weight of the car.

Choosing A Leaf Spring

Leaf springs must be made from a quality material – such as chrome vanadium steel – so that they hold an accurate arch measurement, proper spring rate and load capacity. Multileaf spring ends should be tapered, and there should be Teflon rub blocks between the leaves.

Monoleaf Springs

When choosing the proper leaf spring for your race car you must first determine the weight of the car

The proper monoleaf spring is chosen by car weight and leaf thickness.

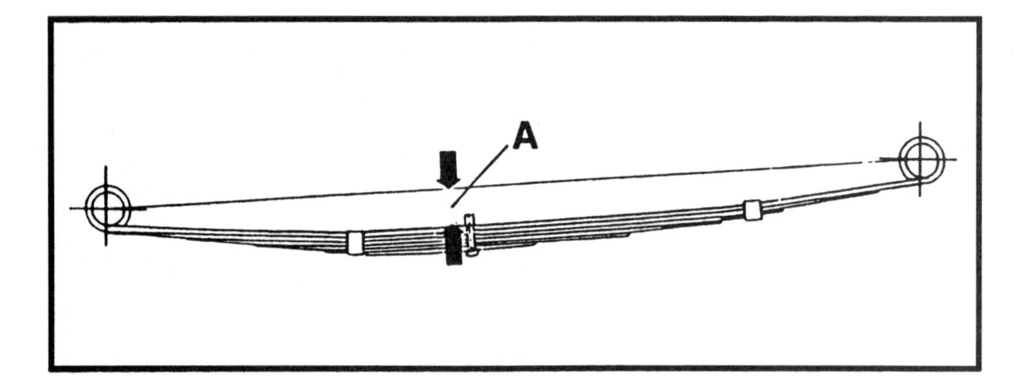

Distance "A" is the leaf spring arch.

and take into consideration how often you want to change leaf springs. Monoleafs seem to work better and hook the car up harder when a thinner spring leaf section is used. The most common monoleaf thicknesses are .323-inch (the thickest), a medium spring at .291-inch, and the thinnest being .262-inch. Obviously the thinner spring will not last as long and must be replaced more often than a thicker one because the thinner section is deformed and twisted more often than a thicker one. But the thinner one will perform better, especially on a dry slick race track. They are more forgiving and cushion wheel spin better.

The best overall monoleaf spring thickness that we have used is the .291-inch. It performs well and doesn't have to be replaced as much as the .262-inch thickness leaf. If I were to suggest a monoleaf, this would be the one. I would also suggest replacing a monoleaf spring every quarter of a season.

Leaf Spring Arch

Leaf springs – both monoleaf and multileaf – come in different arches. The most common would be a 5-inch arch. There are also 6 and 7-inch arch springs available. The different arch heights in springs affect spring rate and tension, depending on the mounting points. A 5-inch spring that is mounted so that there is no tension (compression) on the spring at static ride height (which is the mounting installation we recommend) would be different than if a 7-inch arch spring was mounted in the same manner. The taller spring would be in compression with a 7-inch arch because the rear of the spring would have to be pushed down two inches to line up with the rear mounting points because of the taller arch. This would in turn put more spring rebound force in the chassis. As you can see, a chassis designed for 5-inch arch springs is not going to be the same if you use a 7-inch spring in it. Chassis manufacturers design a chassis to accommodate a certain arch-size spring. Installing a different arch size can create severe changes in the overall chassis geometry and setup.

The leaf spring slider mount positively locates the spring during suspension movement and helps eliminate spring bind.

Leaf Spring Mounting

Shackle mounts and slider mounts are the two different types of mounting for the rear of leaf springs – both monoleaf and multileaf. The shackle mount is the most common and is adjustable up and down. But shackles also have some inherently negative problems. Shackles allow lateral distortion of the leaf spring, which causes bind and makes the spring rate stiffer. The slider mount eliminates this problem.

Slider mounts use a roller bearing and Teflon bushings on the inside. Sliders have a horizontal slot that lets the spring slide back as the leaf spring increases or decreases in length during compression and re-

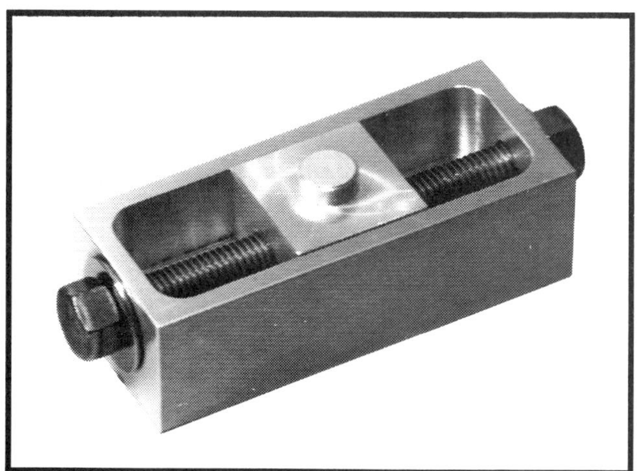

The AFCO adjustable lowering block has a center pin set on a screw. That allows it to be moved forward or backward to lead or trail the right rear tire.

bound. The shackle swings back and forth in an arc motion to accommodate the change in spring length. The slider mount is a better system because it helps to positively locate the spring during suspension movement and helps eliminate spring bind caused by shackle distortion.

Monoleaf springs must be mounted at an angle to the centerline of the race car because they twist and deform laterally during cornering. To achieve this, the rear mounting locations are 5 inches wider apart than the front mounting locations. In other words, each rear mount is moved outward away from the vehicle center line 2.5 inches. This eliminates the spring trying to roll under during extreme side bite conditions. When the springs are angled they resist folding under and lateral deflection during heavy side loading.

Leaf Spring Eye Bushings

Leaf springs should always be mounted in bushings that cushion suspension input loadings, but do not allow forward or lateral displacement of the spring.

It is best to use polyurethane spring eye bushings which are semi-resilient in nature. They are rigid enough to limit spring movement against the mounting bolt, yet resilient enough to absorb vibrations and shock loads. Solid metallic spring eye bushings cause vibrations and shock loads to be transferred to the chassis. This can cause cracked and/or broken frame members.

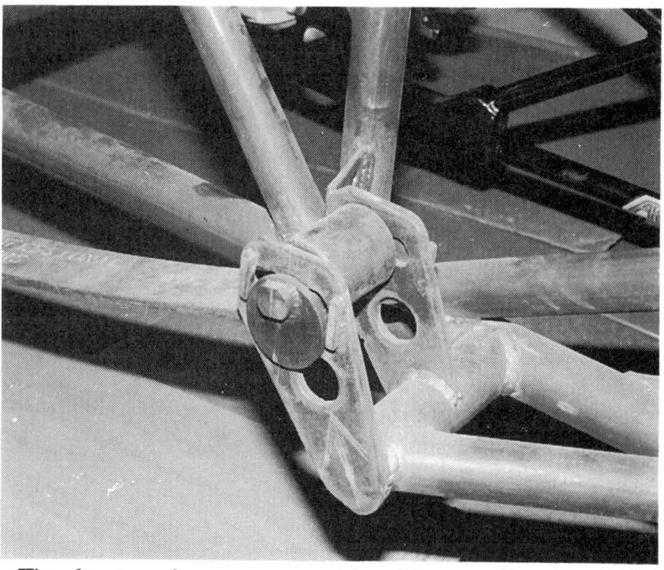

The front spring eye eccentric will move the forward mounting point of a leaf spring 3/8-inch either side of center. Moving the pivot back adds roll oversteer, forward adds roll understeer.

Another type of front leaf spring bushing to consider – for both monoleafs and multileafs – is the AFCO leaf spring pivot bushing. This allows the spring the freedom to move about the front mounting position. The AFCO pivot bushing is designed to replace the bushing in the front spring eye of AFCO and Chrysler-type racing leaf springs. The purpose of it is to relieve any binding or restriction on the movement of the spring when responding to various loadings such as suspension travel, side loads, excessive tire stagger, or other loadings. It allows the front eye of the spring to rotate and twist about the interior ball to keep the spring free.

Slider Lowering Blocks

Adjusting the wheelbase with leaf springs has always been a problem, but now there are lowering blocks which have an adjustable slider mechanism in them to move the axle housing forward or backward to lead or trail the right rear tire. This 1.5-inch tall aluminum lowering block has a center pin that mounts to the housing pad and can move forward and backward on a contained screw mechanism. This part can be purchased from GRT or most race car part suppliers.

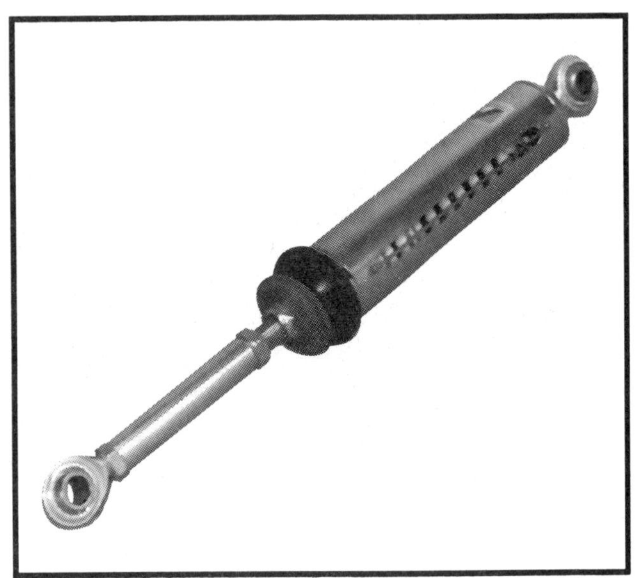

The spring-loaded torque link is a rod connected to a spring or springs contained in a housing. The spring is compressed under acceleration torque to dampen torque reaction at the rear tire contact patches.

3-Point Rear Suspension

The 3-point rear suspension is, in our opinion, the most practical and simplest rear suspension available. We use it on a lot of our sportsman cars and nearly 75 percent of our IMCA modifieds. It uses two lower trailing arms and an upper center link or pull bar to control rear end wrap up. We generally use a spring-loaded center link that angles downhill about 20 degrees. The two lower trailing arms usually mount 3 to 5 inches below the axle center line. The most common center-to-center length for a lower trailing arm is 20 inches. We mount the lower trailing arms at 5 degrees uphill. This uphill angle helps promote better forward bite, but both links being at this angle also creates a slight roll oversteer.

The trailing arms should be fabricated from 1-inch O.D., .156-inch wall steel tubing, and use .75-inch spherical rod end bearings.

The 3-point rear suspension requires a Panhard bar to laterally locate the rear end in the chassis. With this rear suspension, there are not as many adjustments for bite or rear steer as in other systems. IMCA modifieds are really successful with this type of suspension because of its simplicity, so the driver can't dial himself out of the ballpark with adjustments.

A wheelbase adjustment to control rear steer can be made by shortening or lengthening one of the trailing arms. Changing the angle of the center link or pull bar usually makes the largest change in forward bite for this type of suspension. To increase initial forward bite, put more angle in the bar. But the hook-up doesn't last as long with the greater angle. By running less angle the car will hook up more smoothly but not as quickly. Too much upper bar downhill angle can have a negative effect on the chassis under braking. During braking this upper link angle will create pro-squat (rear end rise due to mechanical linkage), which will lighten the rear end and cause wheel hop. This problem can be eliminated by mounting the brakes on separate floaters and not directly on the rear end housing.

Changing spring rates on a spring-loaded upper torque link will have an effect on how the tires are loaded for traction. Softer springs usually promote bite by not jerking the tires loose. Going too soft, on the other hand, will just keep compressing the spring and it will not apply the downforce needed to hook up the car. The opposite occurs when springs which are too stiff are used. Spring rates that are too stiff will cause the tire contact patch to shear and unhook the car. The best possible torque link spring rate recommendation for a 3-point is 700#/" to 900#/" on IMCA cars. Late models usually use 900#/" to 1200#/" springs in the upper torque link. The 3-point suspension is good for sportsman and IMCA cars, but we do not recommend it for late models.

Four Link Suspension

The 4-link or 4-bar system is the suspension of choice for GRT. We have developed and used this suspension for eleven years. The 4-link uses two forward-facing radius rods on each side of the axle housing attached to a birdcage bracket on each side. First let's talk about what a 4-link does.

A four-link provides several advantages over other suspensions. It creates good forward bite because the upper link angles are mounted uphill. Any time a bar going forward is mounted higher on the chassis than it is on the axle, the rear end is trying to go up underneath the chassis and thus the tires are increasingly loaded as the car is being pushed forward. This is because the upward-angled arms are reacting

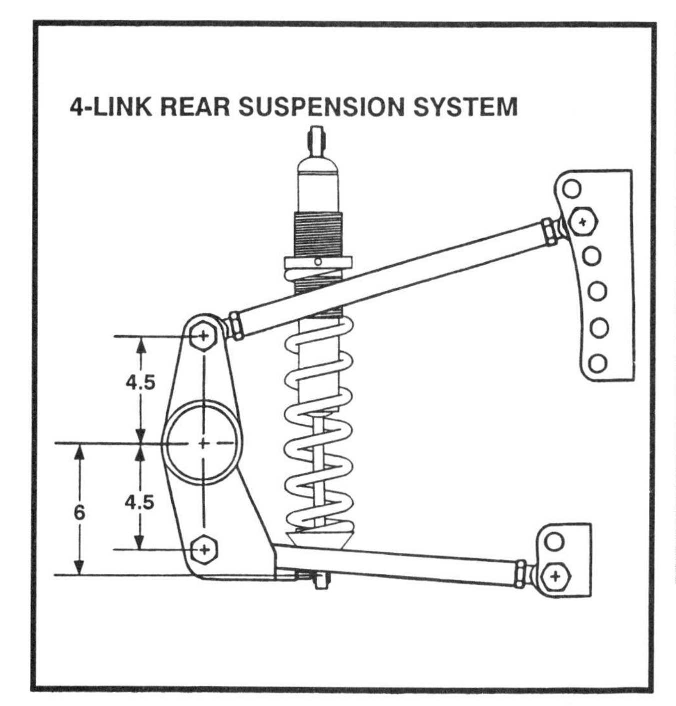

The upper and lower links of a 4-link system attach to a floating birdcage on the axle housing.

against the weight of the chassis which provides more tire loading. This is called axle thrust.

The 4-bar system is great for creating and controlling dynamic rear roll steer. Roll steer is the change in angle of the rear wheels relative to being pointed straight ahead. The right rear is pulled forward or rearward. Dynamic means that the roll steer is not built in, such as pulling the right rear forward with a shorter trailing arm. Dynamic roll steer occurs as the body rolls during cornering due to the arrangement of the linkages that attach the rear end housing to the chassis.

The linkage layout of the 4-bar system creates rear oversteer as the car gains body roll. The right rear wheel is pushed back, and the left rear is pulled forward. This in turn helps the car turn or go through the corner. This really helps a driver to keep the car straight and reduces the chance of the rear of the car hanging out or getting sideways.

The third asset of a 4-bar is the indexing of the floaters (birdcages) during body roll. Indexing is the rotation on the housing tube of the birdcages caused by the differences in angle and length of the upper and lower links. At the right rear, for example, when the suspension compresses during body roll, the coil-over mount at the front of the birdcage rotates forward. This pushes the coil-over mount upward against the spring and shock, resulting in the coil-

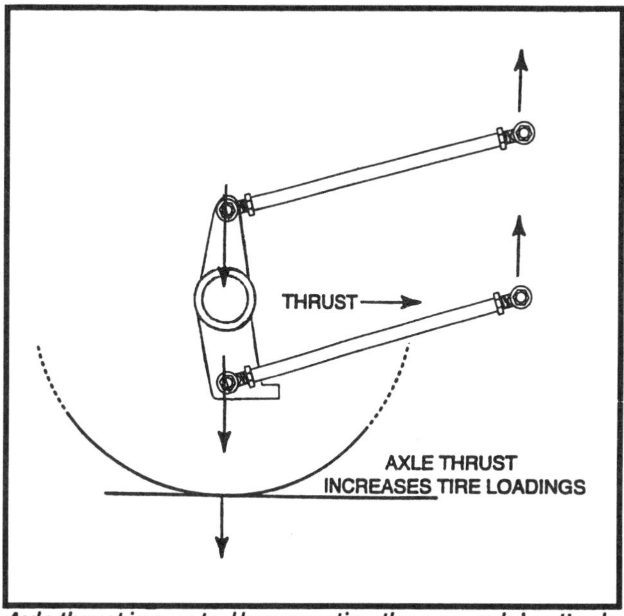

Axle thrust is created by mounting the rear axle's attaching linkages uphill. Drawing courtesy of AFCO Racing Products.

over being compressed from both ends, and stiffens that corner of the car.

Indexing makes the spring actually compress farther than the rear suspension moves. In other words, say the car rolls over 2 inches. With the birdcage indexing, it is compressing the spring and shock at a faster rate than the suspension travel. So in turn you might have 2 inches of roll and get 2.5 inches of actual shock travel. This action is planting the tire into the track surface harder. While this is taking

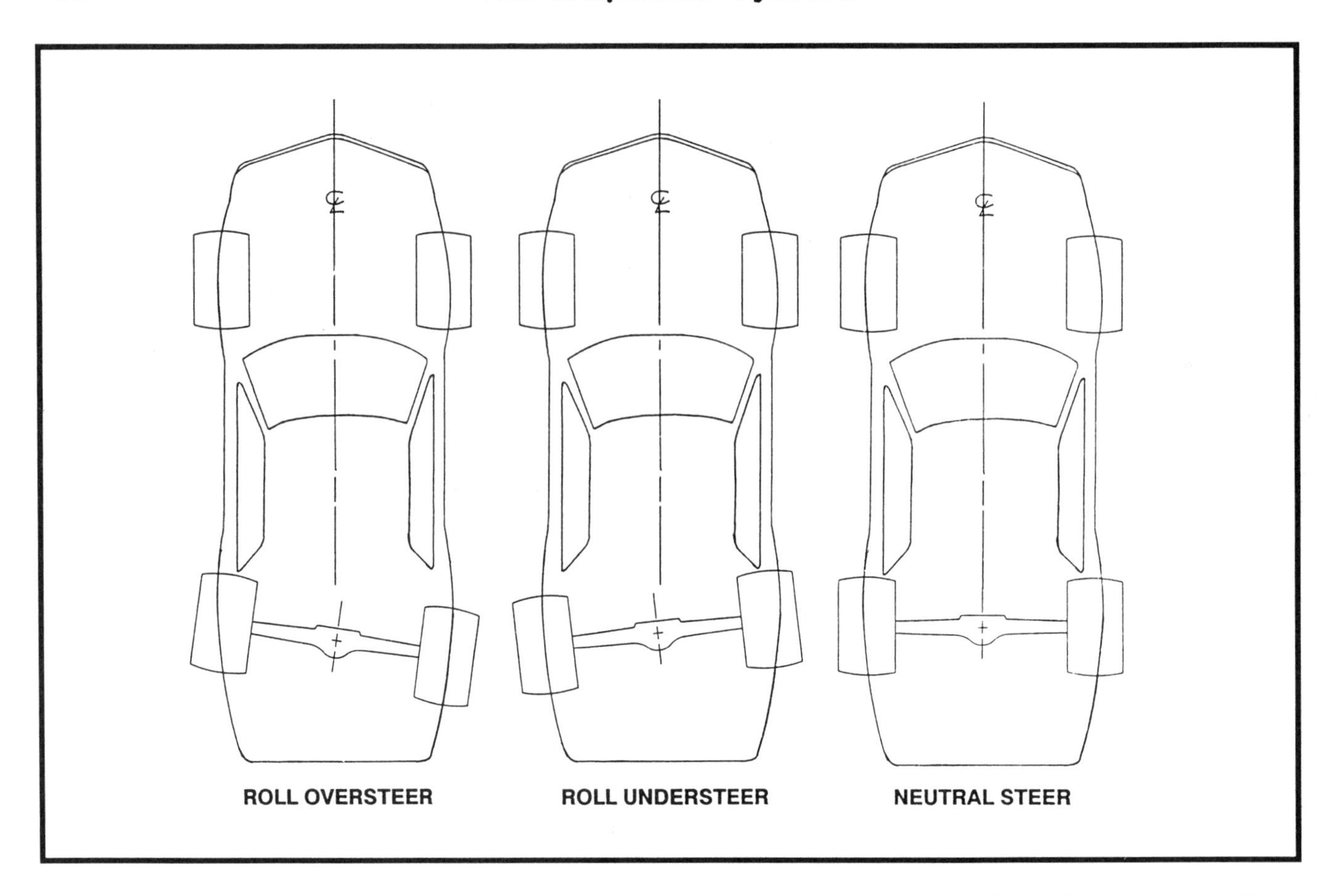

place on the compression side, the rebound has an opposite effect at the left rear. The indexing is rotating the coil-over mount backward which will unload the shock and spring in rebound travel. This reduces crossweight or wedge in the chassis slightly during body roll.

Adjusting The 4-Link Suspension

All of these actions can be altered or changed by adjusting bar angles or lengths. For instance, reducing the angle of the upper link of a 4-bar suspension will take rear steer out and produce less indexing. Reducing the upper angle also softens the rear suspension because of less indexing. Increasing the upper 4-bar angle will do the opposite, but you can go too far (18 degrees uphill should be considered the maximum). If you raise the upper bar too high, the suspension becomes rigid and too stiff. The increased angle does not let the spring and shock do their job and reduces their effectiveness. Ideally, start with the left upper link at 13 degrees and the right upper link at 17 degrees.

The upper links should be at least 17 inches long and be mounted with an uphill angle of 15 to 17 degrees. Usually the left upper link will be mounted with 3 to 5 degrees less angle than the right so the two upper links will be parallel through the corner. If the left upper bar is at the same angle as the right, when the car rolls the right link decreases in angle and the left increases. If the left increases in angle too much, this could create a push condition because the chassis would become too stiff on the left side. So it is best to run the left upper bar at less angle so that both upper links have equal angles at maximum body roll.

The lower 4-bar links are usually 2 inches shorter than the uppers. The reason for this is to reduce the effect of roll oversteer. It is desirable to have a certain amount of roll oversteer to help the car turn freely. But if the bottom links were the same length as the top, there would be too much. Any time you lower the bottom bars or put downward angle in them, it reduces roll oversteer and increases indexing. The most common angle setting for the lower 4-bar links is 5 degrees down. We don't recommend using lower links more than 2 inches shorter than the

4-Link Dynamic Movement

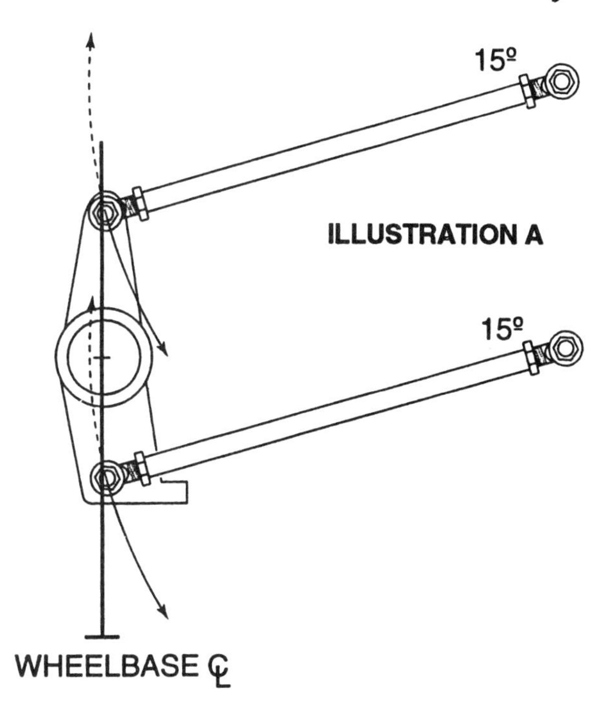

A If the upper and lower links are equal length and mounted at equal angles, during cornering body roll the right side links move equally and push the right side of the axle rearward, while the left side link paths pull the axle forward. This creates roll oversteer. However, the amount of oversteer gained with this configuration is excessive.

B The upper links are still angled upward to take advantage of axle thrust to load the rear tires. But the lower links are angled downward toward the front. This reduces the rearward movement at the bottom of the birdcage on the right rear and left rear, thus reducing the amount of roll oversteer created during chassis roll.

C With a shorter lower link, the arc of the link at the birdcage attachment is sharper, resulting in the right rear birdcage being pulled forward more than with the longer lower link. This further reduces roll oversteer.

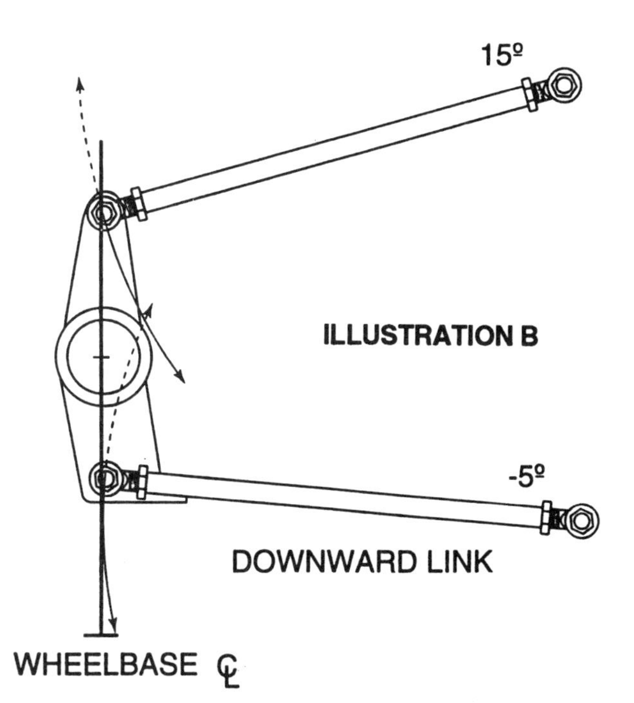

15º

ILLUSTRATION C

-5º

SHORT/ DOWNWARD LINK

WHEELBASE ℄

NOTE: The dotted link path lines represent the right side travel in compression. The solid link path lines represent the left side travel in rebound.

Drawings courtesy of AFCO Racing Products

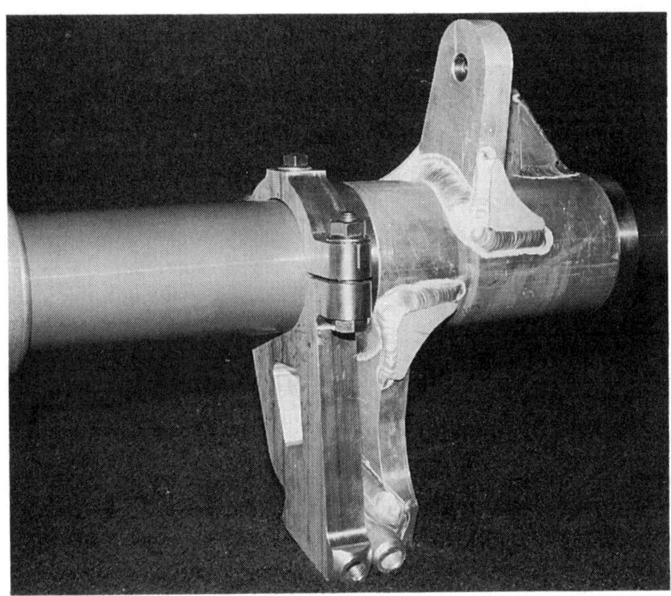

Left rear clamp bracket installed on axle housing.

uppers for most applications because this will create inconsistent handling characteristics.

Study the illustrations for examples of what is happening to a 4-bar suspension under compression and rebound. As you can see, the 4-bar is somewhat complex but very durable and consistent. Once you have determined the best setup for a particular track, the car will function the same lap after lap. There will be no inconsistency. We have also found that the basic general setup of the 4-bar system will get a car close to dialed-in with little need for adjustment for most tracks and conditions. Spring and shock rates are all that is generally required to tune the chassis. And once you have learned the 4-link system and know what changes to make to your car, you will have the advantage over your competition.

What I suggest to you is to study the diagrams shown and build some type of 4-bar scale model that you can play with. This way you can learn what happens when you raise or lower a bar. You can benefit a lot from taking the time to do this.

4-Link With Spring Rod Link

Some racers like to incorporate the use of a spring rod link in the 4-link system. It is usually used as the upper right bar and this helps tighten the car under acceleration. The spring-loaded link lets the rear end go forward which shortens the right side wheelbase and creates understeer. This also takes out indexing of the right rear and softens the right rear spring thus loading the left more which helps the car drive off of the corner harder. A major side effect of the spring rod is handling inconsistency. It does not allow the driver to be as smooth. This type of system has its place but I like the solid rod under most conditions.

4-Link With Clamp Bracket

Taking the shock off of the floater and attaching it on a separate clamp bracket is another common procedure used with 4-bar systems. A clamp bracket is a coil-over mount that bolts solidly to the axle housing and is affected only by axle wrap up and not 4-bar suspension movement. Axle wrap-up is the rotational force of the rear end housing caused by acceleration.

When the coil-over is attached in front of the axle housing to a clamp bracket, the axle wrap-up force loads the coil-over under acceleration. The reaction is an immediate loading of the rear tire at that corner. It lifts upward against the chassis. When used at the left rear, this loading force creates more forward bite and tightens up the chassis at corner exit. Bear in mind the clamp will not rotate any more than the amount the torque arm will travel. But you get immediate reaction under acceleration whereas if the coil-over was mounted on a floater, the spring and shock would not react until body roll occurred. Clamp brackets also soften the rear suspension rate because there is no indexing of the floater. Generally spring rates are increased 25 pounds per inch when running clamp brackets.

Clamp bracket mounting seems to work best on tight-cornered slick tracks. They are generally used on the left rear. I don't recommend using them on the right rear. These type of brackets also unload a chassis rather suddenly under deceleration and can create inconsistency.

There are two different types of clamp brackets in use — a separate clamp bracket and a 3-piece birdcage. The separate clamp bracket bolts onto the housing tube to the inside of the floater bracket.

The 3-piece birdcage uses a normal birdcage with another floating bracket installed to the inside of it, and then a lock-up ring to the inside of that bracket. The coil-over mounts to the inside floater bracket, next to the lock-up ring. When the suspension is to be used as a normal floating 4-link, there is a bushing that attaches between the bottom tabs of the two

The 3-piece birdcage uses a normal birdcage with another floating bracket installed to the inside of it, and then a lock-up ring to the inside of that bracket.

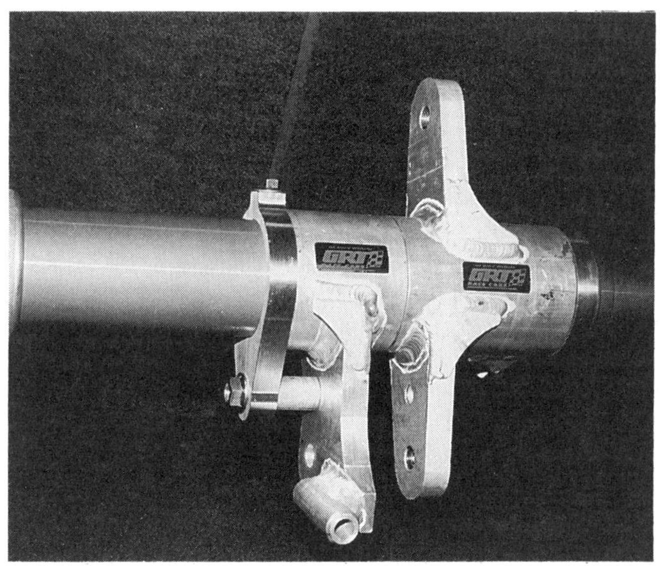

The 3-piece birdcage in the locked position. The bushing is connected between the inside floating bracket and the lock-up ring.

floating brackets, locking them together. When it is desired to switch this suspension to a clamp bracket, the bushing is moved from between the two floating brackets and attached between the inside floating bracket and the lock-up ring. This then locks the inside floating bracket solidly in place, making it function like a clamp bracket instead of a floater bracket. The important feature of this 3-piece bracket is that the coil-over does not have to be dismounted from the floater bracket and reinstalled on the clamp bracket. All that is required is to relocate the bushing from the outside two brackets to the inside two brackets.

When the left rear is equipped with both a floater bracket and a clamp bracket (either separate clamp or 3-piece), it is important to preset the clamp bracket positioning when the car is scaled. If you don't do this, you risk changing the weight setting and corner height at the left rear when you change the coil-over from one to the other at the track.

When the car is being scaled, first get the corner weight set with the coil-over attached to the 4-link floater bracket. Then switch the coil-over to the clamp bracket. The static weight should stay the same. If it does not, adjust the height of the clamp bracket by rotating it until the weight setting matches the amount that was scaled on the left rear with it attached to the floater bracket. This allows you to switch from one bracket to the other at the track and not change weight settings. Install the set screw in

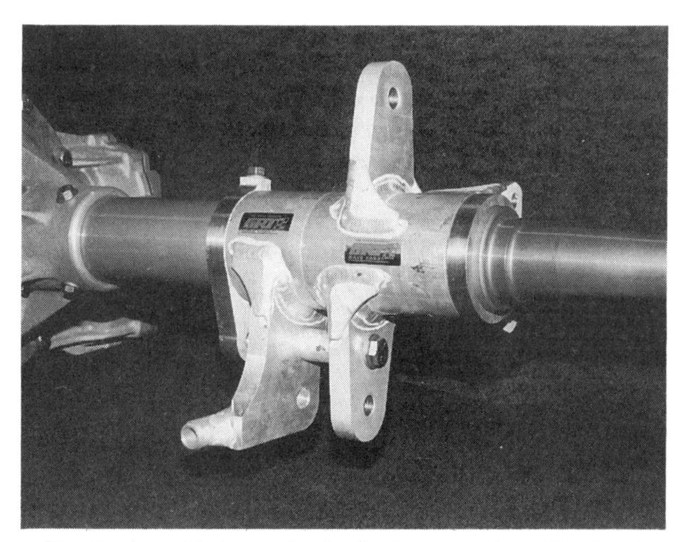

The 3-piece birdcage in the floating position. The bushing is connected between the two floating brackets.

the clamp bracket or lock-up ring with red Loctite, and the bracket is preset.

Other 4-Link Suspension Notes

Brake calipers can be mounted in two different ways with the 4-bar suspension — either on the floater itself, or on separate brake floaters on the axle housing. When a brake is attached to the floater, it will affect the suspension mount through the radius rods under braking. If they are mounted on the housing then all of the brake force is controlled

Brake floaters mount the brake calipers on a separate floating bracket, with braking forces input to the chassis through a radius rod.

through the torque arm shock absorber, the rebound chain, and/or the 6th coil.

If the calipers are on a separate floater with a separate force reaction rod, you can adjust brake force through the chassis without affecting the main suspension.

Our newest chassis design has the brake calipers mounted solidly to the axle housing. We have done this to eliminate any brake reaction forces interfering with the suspension movement. This improves the coil-over indexing by making the reaction smoother. The braking reaction forces are controlled through the torque arm shock. We do not use the brake floater system simply because there is not a big enough advantage to have the extra parts on the car.

Z-Link Or Watt's Link Suspension

This type of rear suspension is similar to a 4-link except that the upper bars run backwards to the rear of the chassis and are the same length as the forward-mounted bars. Because the upper and lower bars are equal in length and parallel to each other, the Z-link suspension produces no roll steer. It also does not produce the same forward bite experienced in a 4-bar suspension which has all the bars running

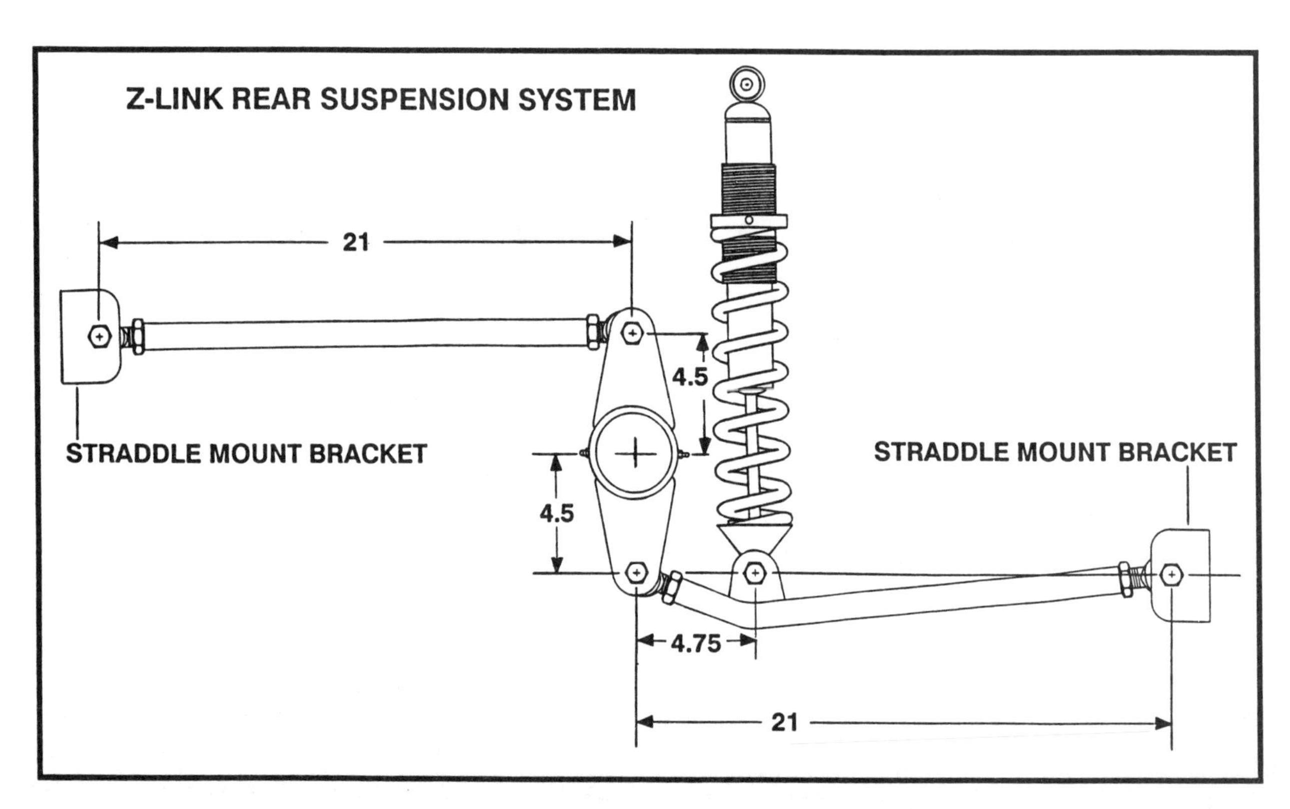

forward, because the Z-link does not produce the forward thrust on the chassis like the angled bars of the 4-link.

Another difference between a Z-link and a 4-link is the mounting point of the rear springs. With a Z-link, the springs are mounted on the front trailing arms directly in front of the rear end housing rather than on a floater. When mounted in this manner there is no indexing of the spring. This produces a real soft type of rear suspension that usually requires stiffer spring rates. This is because there is a greater motion ratio on the spring because of it being mounted further out in front of the housing.

The Watt's linkage type of rear suspension has indexing but no roll steer. This enables you to be able to run the same spring rates as with a standard 4-bar. The Z-link rear suspension doesn't seem to be as tunable and doesn't produce as much forward traction as a 4-bar.

Recommended Rear Springs Rates for a 2,200 lb. Late Model Dirt Car

	LR	RR
Monoleaf C/O	200	175 to 200
3-link	225	200 to 225
4-link	225	225 to 250
Z-link	300 to 325	325 to 350
Watt's link	225	225

Recommended Rear Springs Rates for IMCA Modified Dirt Car

	LR	RR
Monoleaf C/O	150	150 to 200
3-link	200	200 to 225
4-link	175	175 to 200
Z-link	275	275 to 300
Watt's link	175	175 to 200

Recommended Rear Springs Rates for a Sportsman Dirt Car

(2,800 lbs. or above)

	LR	RR
Monoleaf C/O	225	225 to 275
3-link	250	275 to 300
4-link	225	250 to 275
Z-link	350	350 to 400
Watt's link	225	250 to 275

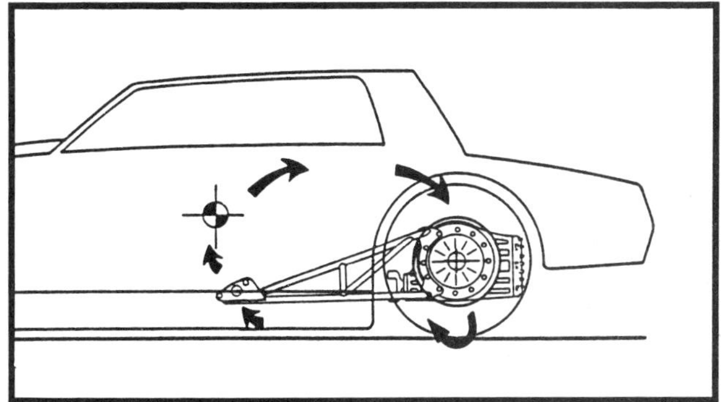

When the rear axle is free to rotate about its own axis in birdcage brackets, the power of this rotating housing can be used as leverage against the weight of the chassis to solidly plant the rear tire contact patches using a torque arm.

Torque Arms

A torque arm is the most common type of torque absorption system used in dirt track stock cars. I've seen all the others tried and when it's all said and done, they are back on the torque arm.

A torque arm is a ladder bar running forward from the rear end housing ranging in length from as short as 30 inches up to 42 inches long. The most common starting place is 38 inches from the center line of the axle. What the torque arm is doing is absorbing axle torque and helping transfer weight to the rear of the car.

The use of floating brackets (floaters or birdcages) on each side of the rear end housing allows the housing to freely rotate without linkage restraints. The rigid torque arm attached to the housing harnesses torque reaction energy and creates a controlled movement in response to acceleration torque. The power of this rotating housing is used as leverage against the weight of the rear of the chassis to help solidly plant the rear tire contact patches against the track surface. This assists in hooking up the car and cushioning the rear tires so that it reduces the chances of shearing the tire contact patch and spinning the tires.

To shelter and protect the rear tires from violent shock reaction, a torque absorbing device is placed between the forward end of the torque arm and the chassis. A coil-over unit (called a fifth-coil) is placed into the torque arm system and allows the downforce to be more gently applied at the rear tire contact patches as the throttle is applied. I like to use a 73 series shock (7-inch stroke, 3 valve code) on the fifth

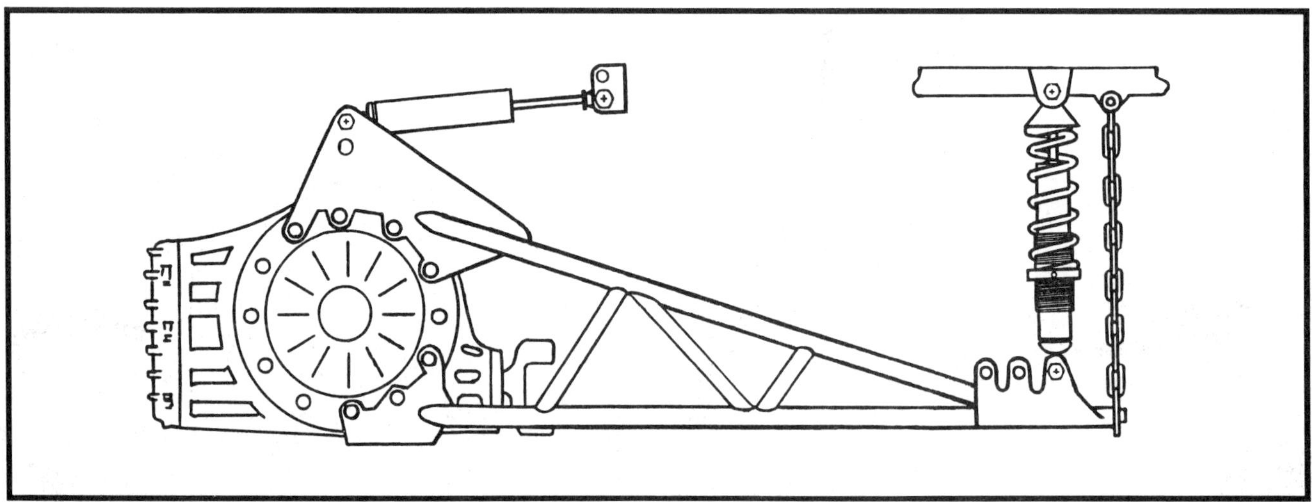

The fifth coil-over torque arm system shelters the rear tire contact patches by placing a spring and shock absorber between the torque arm and the chassis. An axle damper shock mounted at 5 degrees uphill above the rear end housing helps to control braking reaction forces on the arm.

coil and a spring rate of 250 to 350#/", depending on track conditions. Softer springs seems to work best on extremely dry tracks and stiffer springs work better on heavy or "hooked up" tracks.

A better spring that works best on a variety of track conditions and surfaces is the progressive rate spring. They make these springs in soft, medium, and stiff ratings. These springs start off fairly soft under compression and progress to a stiffer rate the more the torque arm travels. This has an advantage over the standard spring in that it cushions the initial force at slower speeds and, as the car gains speed, the torque arm spring increases in rate to keep the car hooked up down the straightaway.

Using too stiff of a spring rate initially will break the tires loose or shear the contact patch. This explains why the progressive spring works so well. We have also found that the heavy progressive rate spring works best under most track conditions. The heavy progressive starts at the same rate as the softer one but increases in rate more rapidly. Usually the rates go from 185 to 200#/" on the first inch to 500#/" on the third inch of compression.

Torque arms usually work best traveling 3 to 3.5 inches. We always preload our 5th coils about .5-inch. A short torque arm usually works best with a stiffer spring and hooks the car up sooner, but this reaction doesn't last as long. This works best on a real short stop-and-go type of race track. Longer torque arms don't hook the tires up quite as quickly but have a longer, smoother hook up. Bear in mind

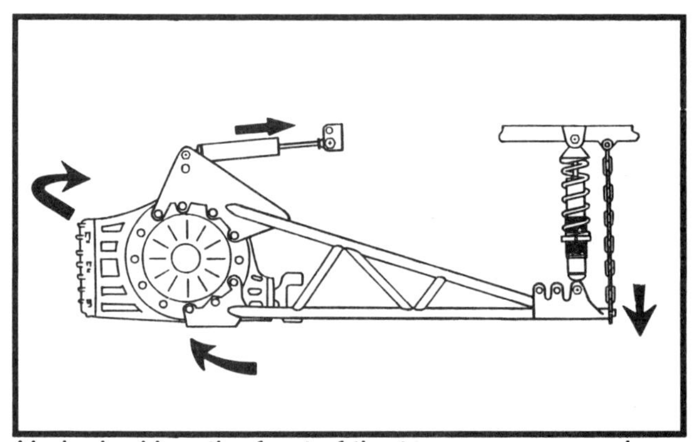

Under braking, the front of the torque arm goes down. If these forces are not reacted, they will lift the rear wheels off the ground. The axle damper shock is used to resist these forces. A restraining strap, such as a length of chain, limits the downward movement of the front of the torque arm.

that if a torque arm is too long and you are running on a big sweeping long-turned race track, it has a tendency to lift or carry the front wheels and create a push. So don't create problems by using too long of a torque arm. Stay in the 38-inch range.

We recommend using a rebound chain on the front of the torque arm for stopping the downward movement of the torque arm during braking. Our chains are the horseshoe type with a double chain that lets the mounting location be in the same exact location as the fifth coil shock.

With a 4-link system, use a brace that connects to the chassis and supports the end of the arm. This keeps the arm straight no matter where the rear end points.

The other type of lateral support system solidly bolts the rear attachment points of the torque arm to the axle housing center section. It uses a bracing bar that attaches to the front of the torque arm and triangulates back to the rear end housing tube

Torque Arm Bracing

Torque arms should be braced well to keep them from moving from side to side and loading and unloading the spring. A lateral movement of the torque arm could also cause it to hit the driveshaft. There are two different types of braces that can be used – a solid mounting or a floating mount. This is an important area and the choice of bracing system depends on what type of suspension you use.

With a 4-link and the roll steer associated with it, you need a brace that connects to the chassis and supports the end of the arm. This keeps the arm straight no matter where the rear end points. Therefore it keeps the shock and spring directly under the mounting point in the chassis and you have maximum performance at all times. This system requires the torque arm mounting points to float on rod ends at the differential center section connection.

With the floating brace system, the rear attachment points of the torque arm are bolted through rod end bearings to the center section. This mount keeps the torque arm rigid in vertical movement, but allows it freedom to move laterally. At the front of the torque arm, a tubing brace is bolted from the arm to the frame on the right side. This brace has a rod end bearing on each end. What this system does is force the torque arm to move in a straight up and down motion while allowing the rear end housing to pivot to accommodate roll steer.

The other type of lateral support system solidly bolts the rear attachment points of the torque arm to the axle housing center section. It uses a bracing bar that attaches to the front of the torque arm and triangulates back to the rear end housing tube on the right side. With this system, when there is roll steer the torque arm follows the path of the rear end housing and may swing as much as 2 inches left to right, depending on how much roll steer is achieved. This puts the fifth-coil unit at an angle during roll steer. Either system works, but the floating type of support is more consistent.

Sixth-Coil Torque Unit

Sixth-coils are sometimes used on torque arms instead of a rebound chain. This system cushions the rebound at the end of the arm instead of completely stopping the downward motion like a chain. This generally will tighten a car (create understeer) at corner entry.

An area where a sixth-coil becomes even more critical is when the brake caliper brackets are mounted on the housing and the brake force is fed through the torque arm instead of through radius rods from brake floaters. The sixth-coil then has more of an effect on corner entry to keep the car

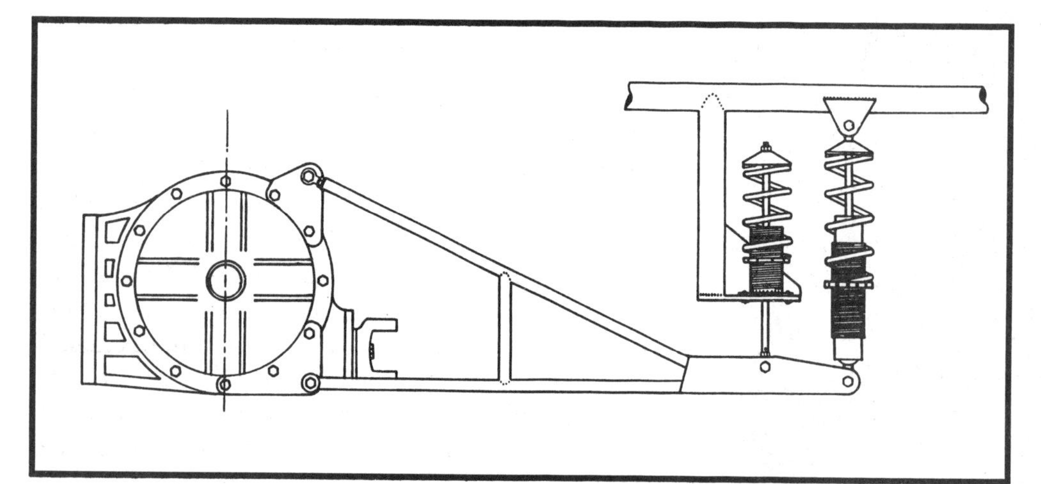

Sixth-coils are sometimes used on torque arms instead of a rebound chain. This system cushions the rebound at the end of the arm instead of completely stopping the downward motion like a chain.

tight. An inherently loose race car could use the sixth-coil set up. Or, it would work well for a driver that likes his car extremely tight on corner entry. Our particular cars work well with a chain with or without the brakes mounted to the axle housing. These choices all depend on the driver's style and the way the chassis is set up.

Axle Dampers

An axle damper or 90/10 shock absorber is placed above the rear end housing and is angled uphill to the front of the car (usually 5 degrees). This is used with any torque arm system. The 90/10 shock (which means 90% of the rate is absorbed on compression, 10% on rebound) cushions or slows down the unwrapping of the rear end housing under braking before and while the chain or 6th-coil is being used. If you are using a rebound chain and the car is under braking, by the time the chain bottoms out the rebound shock has kept any severe impact from happening. The driver can't ever feel the rebound chain stop. If the track is extremely loose on entry, we suggest using two 90/10 shocks. They need to be mounted parallel and angled uphill 5 degree for best performance.

When mounting a 90/10 shock, if the shock is mounted to the left of vehicle center line on the rear end, the car will have a tendency to be looser on turn entry. If the 90/10 shock is mounted to the right side of the center line on the rear end, it would be tighter. The reason for this is that the right rear tire, with the most load under braking and corner entry, will try to pull the car in that direction. This is similar

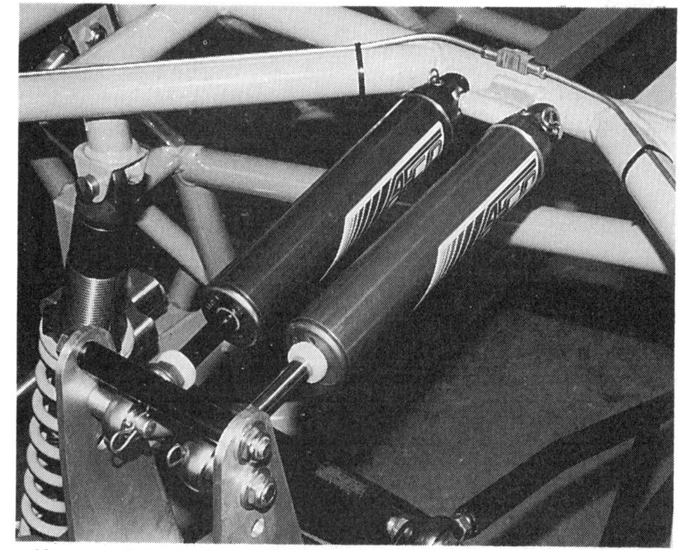

If a track is extremely loose on entry, use two 90/10 shocks. They need to be mounted parallel and angled uphill 5 degrees for best performance.

in reaction to a bulldozer that stops one of its tracks and makes the bulldozer turn left or right depending on which track is stopped. Stop the left one and it turns left; stop the right one and it turns right. This same concept happens on a race car. With more traction at the right rear tire under braking, the car wants to turn right. If the left rear has better traction under braking, the car will be looser.

The Spring-Loaded Torque Link

As discussed earlier, the spring-loaded third link is a type of torque absorption device with which we have had little luck on dirt late models. Third links are erratic and hard to control. I firmly believe there are too many linkages involved with a third link and

This car uses a spring loaded torque link with an axle damper shock mounted above it.

a 4-bar rear suspension combination, and they bind up somewhere in suspension movement.

However, there are classes where rules prohibit the use of torque arms, so they can use the spring-loaded third link as an effective torque absorbing device. The 3-point rear suspension is a good place for a spring loaded torque link. Spring rates have to be extremely stiff on these because the mounting point is so close to the centerline of the rear end. The spring-loaded third link works well on an IMCA modified because the modifieds use a smaller diameter and narrower tire, and thus they don't produce as much force on the third link spring.

Panhard Bars

Basically there are two types of Panhard bars — a J-bar and a straight bar. Both are used to accomplish the same thing — to brace the body of the car against the rear end housing so that there is no side shift of the axle during cornering. They are a lateral locating device. Both types are available as a solid bar or they can be cushioned with springs or rubber biscuits. The Panhard bar is an important linkage, because it locates the rear roll center (both the height and the lateral location), and it can create leverage, depending on its mounting angle and location, to influence traction at the rear tires during cornering.

The J-bar design allows you to maintain a longer length Panhard bar and also have plenty of adjustment even with the driveshaft being in the way. The J-bar goes over the driveshaft and makes a downward bend toward the ground. This creates the angle that is desired when mounting the bar. On the other hand, a straight bar cannot be mounted as low because of driveshaft interference. Panhard bars are mounted to the front of the rear end housing when a quick change rear end is used because the design

(Right) Three different styles of a Panhard bar: (top) the rubber cushioned bar, (center) the spring cushioned bar, and (lower) the solid bar.

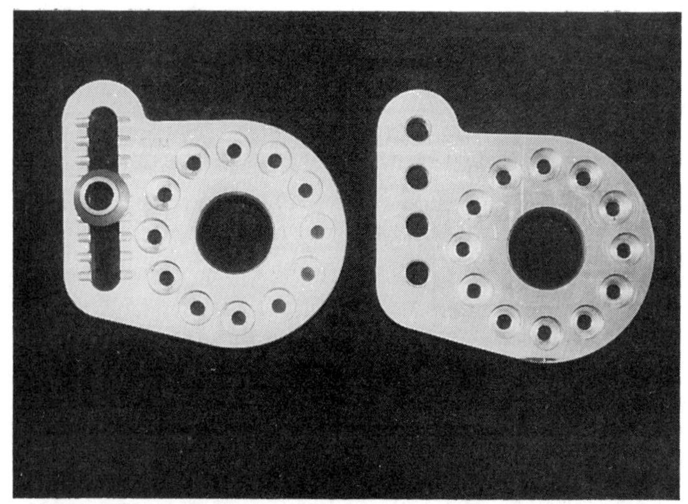

Two different styles of quick change Panhard bar mounting brackets. The one at the left has a serrated slide adjustment in 1/2-inch increments. The one at the right uses conventional threaded bolt holes.

of a bar to mount around the quick change protrusion would be too cumbersome and heavy.

J-bars are extremely popular and very adjustable. A normal mounting strategy would be to attach the bar to the left side on the chassis and the right side to a bracket on the rear end housing (pinion-mounted Panhard bar brackets are the most common). This creates a downward angle from the chassis mount to the rear end, with a difference of approximately 3 inches in vertical height from the chassis mount down to the rear end mount.

The Panhard bar is mounted to the chassis on the left side by means of a clamp bracket. The bracket can be easily moved up or down in small increments to adjust the roll center height.

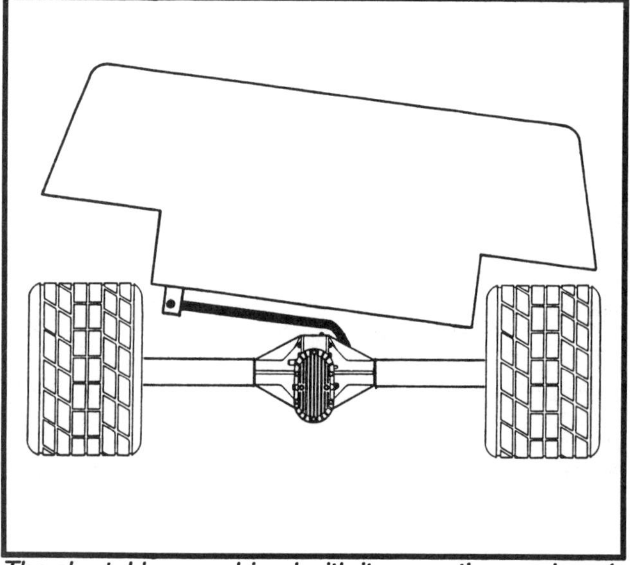

The short J-bar combined with its mounting angle pole vaults the body over the roll axis of the car. The overturning weight is transferred to the pivot point connection on the axle, adding more down force to the right rear tire.

The J-bar is normally 19 inches long. This is our standard length and it works well with most track conditions. When the bar has 3 inches of angle in it, this produces mechanical roll in the car and actually plants the right rear tire in the ground. This is extremely important on a dry track because a dirt car needs roll to maintain side bite. The basic rule of thumb here is, the more angle in the Panhard bar mount, the more roll that will be produced. And the shorter that the bar is, the more radical the reaction at the right rear tire. I've seen cars have so much angle in the Panhard bar and be so short that they will actually carry the left front wheel all the way down the straightaway. Now that's leverage pushing on the right rear!

One of the major problems causing the radical reaction is bump steer caused by the short length of the J-bar. This type of setup (short J-bar plus radical mounting angle) is not recommended for inexperienced drivers because it makes a car very hard to drive.

It is important to understand how the mechanical roll is produced with a left side-mounted Panhard bar. When a car enters a left hand corner and starts to go through it, the car's body tries to swing over the roll axis and shift all of the weight to the right. At the same time this is going on, the rear end of the car is trying to go up under the car or to the left, the

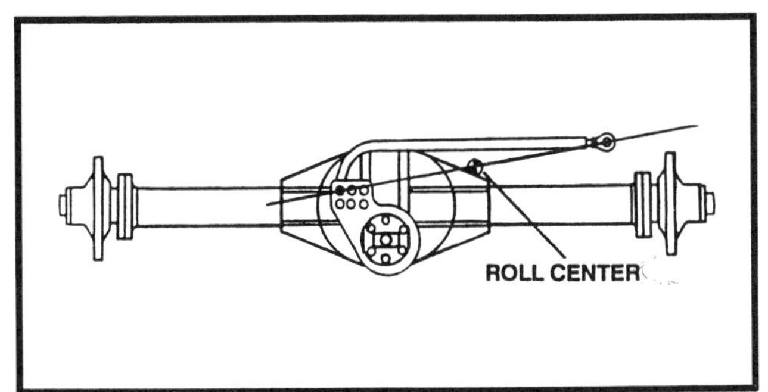

The Panhard bar should be mounted from the rear end housing to the chassis on the left, and have a downhill angle between the chassis mount and the rear end mount.

opposite direction of the body. Since the rear end is connected to the body by a Panhard bar and the left side mounting point is on the frame and is higher than the rear end mounting point, the body is continually rising up trying to roll and the rear end is moving to the left. The body in essence is trying to pole vault over the rear end. Now you can see why more angle and a shorter bar could get out of hand real quick. But this is the reason the left side mounting system works so well on a dirt car. You must meet a happy medium between mechanical roll and natural roll.

A revised system we have developed uses a longer Panhard bar. It still provides the roll required but is not nearly as radical. We are currently using a 24-inch long Panhard bar with extra 5-inch length being added to the left side of the chassis through a revised mounting location. This longer Panhard bar also benefits weight transfer back to the left rear due to the fact it is not trying to hold up the chassis with the pole vault effect. With a short bar, the mechanical force will hold up the left side of the car and not allow weight transfer back down on the left rear corner. The longer bar makes the weight transfer back onto the left rear much easier.

Sometimes cars are so tight at turn entry that you need a straight bar which is level and is mounted at axle tube height. This position is good for a hooked up or rubbered down race track where getting traction is not a problem.

Always remember that no matter where the Panhard bar location might be, if it is on the left, make sure it runs downhill to the rear end housing mount. A Panhard bar on the left with angle in it will produce side bite. You must determine how much you need for your application. The higher you mount a Panhard bar, the freer the car. The lower you go, the tighter the car. This is because a higher mounting produces a higher rear roll center, whereas the lower mounting produces a lower rear roll center. I like to have multiple hole brackets or slider brackets on the frame and the pinion-mount bracket in order to easily adjust the Panhard bar angle.

Left Vs. Right Mounting

We have tried mounting Panhard bars to the right and we can't make it work properly. This seems to be more of an asphalt type set up and works well with pavement cars. Our cars are just too loose with this set-up and we can't achieve any of the mechanical roll needed that you can get from a left-mounted bar.

Cushioned J-Bars

As far as using a spring or rubber-cushioned J-bar, in theory this will absorb some of the side load shock and help keep the car from sliding the contact patch during hard cornering. I can say I have never seen lap times improve with the use of a cushioned bar. I think this is all mental and if you feel like it will benefit your program, use it. If I was going to make a choice of which one to run, it would have to be the rubber type. They are lighter and don't have as much rebound effect.

Chapter
9

Shock Absorbers

Shock absorbers are nothing more than velocity sensitive, heat dissipating, hydraulic devices. The shock absorber offers resistance to movement of the suspension by forcing hydraulic oil through a series of valves and openings inside the shock.

Shock absorbers affect the handling of a race car as much as any other suspension component. Shocks are one of the most misunderstood and most overlooked aspects of chassis tuning. Most racers depend on someone else's recommendations when choosing shocks for their own race car. If the recommended shocks are incorrect, the racer ends up adjusting his chassis around the wrong shocks while trying to correct his handling problems. The end result is the racer turning in a mediocre performance.

We feel, however, that if a racer understands how shocks work and how they affect the car, he can use this to gain the advantage over his competition.

Twin-Tube, Low Gas Pressure Shocks

There are two different types of shock absorbers used in racing vehicles — twin-tube, low gas pressure shocks, and mono-tube, high gas pressure shocks. The twin tube design is most often used in dirt racing.

Twin-tube shocks are built with two different tubes. The pressure tube, or center tube, functions as the cylinder base for the piston. The reserve tube, or outer tube, provides the outer wall for the hydraulic fluid. This type of design helps to keep dents and dings from detracting from the shock's performance. The pressure tube is sealed along with the compression valve. The compression valve regulates the force needed to resist the compression of the shock. A series of valves and parts in the piston regulate the force when the shock is in rebound.

Mono-Tube High Gas Pressure Shocks

Mono-tube, high gas pressure shocks are designed differently from twin-tube shocks. They have a single

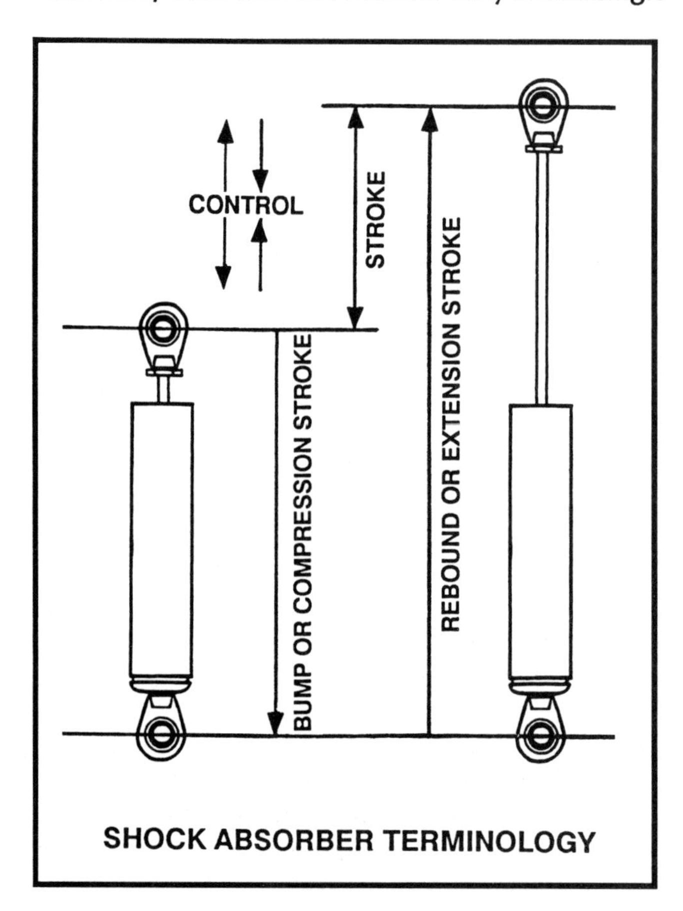

SHOCK ABSORBER TERMINOLOGY

tube, they use deflective discs for valving, and they have a high pressure nitrogen gas chamber which is used to resist oil cavitation (foaming). The mono-tube functions as the inner bore for the piston, as well as the outer shell of the shock body. However, when the shock body becomes dented, the piston movement is hampered, thus resulting in shock failure.

A positive aspect of the mono-tube is that the design allows a larger diameter piston to be used since the bore is larger. This design also requires a second floating piston that is located between the shock oil and the pressurized gas. The gas is what creates enough force to control oil cavitation.

My choice of shock type would have to be the twin tube, low gas pressure shock. We use AFCO shocks in the majority of the race cars we build, but the other shock manufacturers provide excellent products as well.

By using stiffer rebound control shocks on the left rear and the left front, the dynamic weight unloading process from the inside tires to the outside tires slows. Consequently, the left side tires remain loaded further into the corner which helps to turn the car. Photo by Fred Patton

How Shocks Influence Handling

In twin tube shocks, there are a series of valves and orifices. They operate at various pressure levels to resist movement. This staged valving is necessary because the shock resistance required to control the race car suspension when it goes over a severe bump (referred to as high speed control) is much greater than the resistance needed to control body roll or the suspension movement caused by small bumps (referred to as low or medium speed control). Shock resistance at low, medium, and high piston speeds must be matched to the needs of the race car.

In the simplest terms, racing shocks perform two functions:

1. When bumps and ruts are encountered, shocks keep the chassis settled and the tires in compliance with the race track. Without shocks, the chassis would pitch, roll, and bounce violently whenever the race car encounters bumps and ruts. The tires could lose contact with the track surface.

2. Shocks help control the rate of chassis roll and pitch caused by dynamic weight transfer. Whenever a race car accelerates, decelerates, or corners, the chassis will pitch or roll (due to weight transfer). Without shocks, body roll and pitch would be violent and the chassis would not be stable.

Shock control at low piston speeds affects how the race car handles through the corners, while medium and high speed control affects how the race car handles whenever it encounters bumps and ruts.

Rebound control is a shock's resistance to extending and is specified at a given piston speed. The amount of rebound control developed by a shock will determine, generally, how quickly the tire is unloaded during dynamic weight transfer, and how quickly the suspension "rebounds" or returns to its original position, after the spring has been compressed.

Compression, or bump control, is a shock's resistance to compression and is specified at a given piston speed. Compression control will determine, generally, how quickly the tire is loaded during dynamic weight transfer and how the suspension will react whenever a bump is initially contacted.

The stiffness of the shock absorbers used on a race car has a profound effect on the rate at which weight transfer affects the loads on the tires. Because of this, shocks are a very important factor when it comes to handling. Basically, soft shocks allow weight transfer to affect tire loadings more quickly than stiff shocks.

The traction capability of a tire determines that tire's influence on the race car. Traction capability is greatly affected by the load put onto the tire.

The balance of traction between the left side and right side tires determines to a great extent how a car will handle while decelerating through the corner. For example, a race car will tend to push whenever

Shock Absorber Part Number Cross Reference

(All Numbers Are For Steel Body Shocks)

AFCO	Carrera	PRO
1073	3173	7300
—	3172/3	—
—	3176/3	—
1074	3174	7400
—	3173/4	—
1074-6	3176/4	7460
1075	3175	7500
1075-3	3173/5	7530
1075-6	—	—
—	3177/5	—
1076	3176	7600
1076-2	3172/6	—
	—	7630
1076-4	3174/6	—
1077	3177	7700
1078	3178	7800
1079-1*	3171/9*	7910*
—	3174/9	—
1093	3193	9300
1093-5	3195/3	—
—	—	9380
1094	3194	9400
1095	3195	9500
1095-6	3196/5	9560
1096	3196	9600
1097	3197	9700
1098	—	9800

Notes:

1. On split valving shocks, AFCO and PRO specify the compression valving as the first number and the rebound valving as the second number. For example, a 1076-4 AFCO shock has a 6 valving on compression and a 4 valving on rebound. A 7630 PRO shock has a 6 valving on compression and a 3 valving on rebound. Carrera specifies their numbers just the opposite — rebound first and compression second.
2. To convert AFCO steel body shock part numbers to aluminum body shocks, instead of 10 at the beginning of the part number, add 11 for smooth body shock, or add 13 for threaded body shock.
3. To convert Carrera steel body shock part numbers to aluminum body shocks, instead of 31 at the beginning of the part number, add 71 for smooth body shock, or add 61 for threaded body shock. To convert to steel body shocks with weld-on bearings (cannot be used for coil-overs), add 21 in place of 31.
4. To convert PRO steel body shock part numbers to aluminum body shocks, add A at the beginning of the part number and drop the last 0 for smooth body shocks, or add AC at the beginning of the part number and drop the last 0 for threaded body aluminum shocks.

the left side tires do not have enough influence in stopping the car. This happens when the right side tires are slowing the vehicle more than the left, so the vehicle tends to go to the right. By using stiffer shocks (especially a stiffer rebound control on the left rear, and to a lesser degree, a stiffer rebound control on the left front), the dynamic weight unloading process from the inside tires to the outside tires slows. Consequently, the left side tires remain loaded further into the corner which helps to turn the car.

When making this adjustment, consider using the appropriate "split valve" shocks so the compression control of the left side shocks is not increased. This change should allow the chassis to roll back onto the left side tires more easily during corner exit.

Also, the opposite of the above example holds true. Softer extending left side shocks, especially the left rear, will cause the left side tires to unload sooner. The balance of traction between the left and right side tires moves toward the right tires more quickly and the chassis becomes tighter on corner entry.

During acceleration, the balance of traction between the rear tires can be adjusted with shocks also. A softer left rear shock (especially compression) will quicken the weight transfer effect to the left rear tire during acceleration. The result is a left rear tire that has added influence, initially, in accelerating the race car off the corner. A race car will tend to be tight off the corner whenever the balance of traction between the rear tires favors the left.

Forward traction can be enhanced by softening the extension control of the front shocks. This enhances the front to rear weight transfer process and helps to load the rear tires for improved forward traction.

Remember, shocks are a compromise like any other suspension component. Be careful when using split valve shocks with soft rebound controls so that the handling over bumps and ruts does not suffer.

There really is no mystery to shock function and tuning. However, there are complexities and qualities that need to be considered when choosing shocks for a specific application. By keeping this basic information in mind you should be able to install the correct shocks for each situation when troubleshooting handling problems. This should also enable you to have the confidence to make changes with fairly good expectations of results.

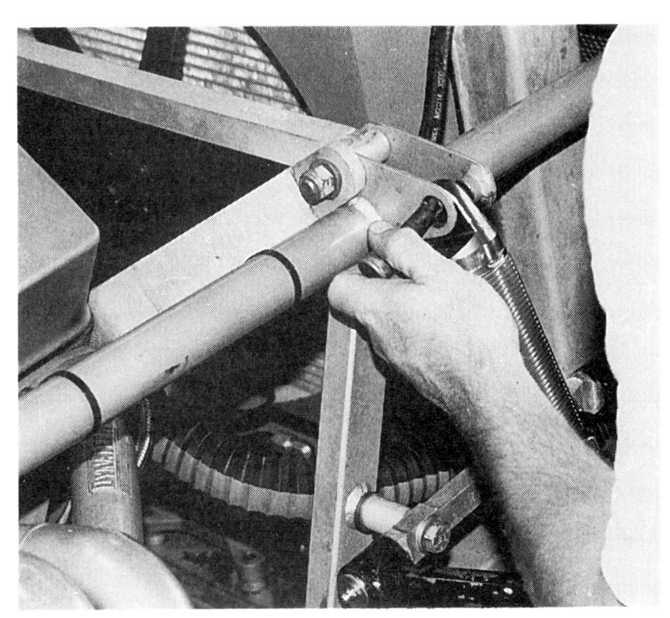

Shocks should be mounted solidly to the lower A-arm on front suspensions as close to the wheel as possible, and should be mounted securely with a bolt or pin.

Rear shocks should be mounted 7.5 to 7.75 inches in from the back edge of the brake rotor and angled inward 15 to 18 degrees.

Above all, remember that chassis tuning is a compromise and shocks, though a very important part of the set-up, are still only a part.

Without a basic understanding of the characteristics of shock absorbers it would be very difficult to correctly tune a chassis. Keep the following in mind for proper chassis tuning:

1. As the piston speed of a shock increases, the shock gets stiffer.
2. Large bumps hit at high speeds cause the highest piston velocities, and the highest shock resistance, to occur.
3. The low speed resistance of a shock absorber controls the rate of body roll and pitch, and also how quickly a tire is loaded and unloaded during dynamic weight transfer.
4. Generally, soft shocks will cause a tire to become loaded or unloaded (due to dynamic weight transfer) more quickly than stiff shocks.

Proper Shock Absorber Mounting

Shocks should be mounted solidly to the lower A-arm on front suspensions as close to the wheel as possible. They should be mounted securely with a bolt or pin. I have seen various shock angles used and our current design left front angle is 15 degrees, while the right front angle is 18 degrees. Both rears are angled inward 18 degrees and 8 degrees forward.

A good bottom mounting point for rear shocks would be 7.5 to 7.75 inches in from the back edge of the brake rotor. Make sure that the shocks are mounted securely and that the rod end has plenty of clearance for all of the movement that will be associated with wheel travel.

When mounting shocks, make sure that the center-to-center installed distance prevents the shock from bottoming out on compression or rebound. We use 7-inch stroke shocks on the front and 9-inch stroke shocks on the rear. They need to be mounted so that when the car is at normal ride height, the stroke travel is centered. A good practice is to allow for a little more stroke travel on compression. A 7-inch stroke shock should have 3.75 inches travel for compression, and 3.25 inches travel on rebound.

Another area of concern is how the mounting angle affects motion ratio. Usually the more angle there is, the lower the motion ratio will be. This does not change the damping force, but it slows the momentum ratio down, therefore the shock piston does not provide as much shock control. Obviously, you want maximum control, so don't use any severe mounting angles on the shocks. If you choose to experiment with shock mounting angles, remember that the closer to the wheel and the closer to vertical

A shock absorber should be mounted on top of a lower A-arm (as shown at right) instead of to the side of the A-arm (left). When the shock as mounted on the side, it is not close enough to the wheel to provide effective control, and it can provide clearance problems with the brake caliper.

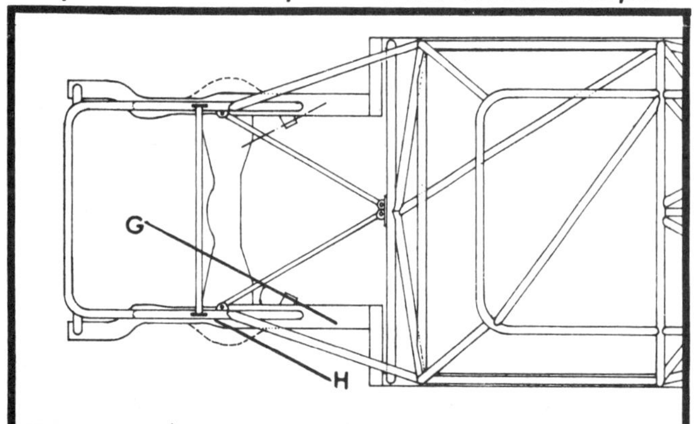

Shock absorbers must be mounted in the same plane as the swing angle of the lower A-arm. Here, "G" represents the swig plane of the lower A-arm while "H" is the proper mounting plane for the shock absorber. If the planes are not parallel, suspension bind will occur.

a shock is mounted, the more wheel control it will have.

Front shock mounting angles should be the same as the swing angle of the lower A-arm.

Matching Shocks To Track Conditions

When you choose a shock valving, make sure you consider the surface to be run on. Too stiff of a valving will more or less let a wheel skate over bumps, and the tire won't stay on the ground, thus loosing traction and speed. The same goes for a shock that does not have stiff enough valving. The wheel will bounce at an excessive rate and not stay in contact with the surface. Again it looses traction and control.

Let me give you some examples of which shocks we normally run on our late models. A standard set-up uses a 75 valving shock on both the left front and right front that are close to 50/50 ratio. Our rear shocks use 94 valving shocks with a 50/50 ratio at each corner. This particular shock combination works well on most track conditions that are fairly smooth, and flat to semi-banked. I like to see a racer use these shocks until the track becomes rough, has a cushion, is extremely high banked, or is extremely flat and dry slick with slow corner speeds.

If the track is high banked and/or has a cushion, we generally go to a 76 right front and a 95 right rear shock. These are stiffer valved shocks, and they help stabilize the car for these type of track conditions.

An extremely high speed momentum track that is real smooth might require a split valving shock on the right front, such as a 76/4 (6 valving compression, 4 rebound) or 76/2 (6 valving compression, 2 rebound), depending on how smooth the track is. On the other hand, a real slow flat track that is slick

A 93/5 split valving shock installed at the left rear lets the weight transfer quicker to the rear under acceleration. At the same time, the stiffer rebound valving helps to hold the left rear on the track surface at corner entry.

could use a 75/3 right front shock. These shocks allow weight to transfer very quickly as the car accelerates off the turns. The soft rebound rate at the right front transfers weight onto the left rear. This helps promote forward bite.

This is the shock dyno installation at AFCO Racing Products.

Another split valving shock we use is a 93/5, installed at the left rear. This is a 3 valving on compression and a 5 on rebound. The 3 valving lets the weight transfer quicker to the rear because it has less compression resistance, and at the same time the 5 valving offers a stiffer rebound, helping to hold the left rear on the track surface. This also helps the car turn in the middle of the corner. Sometimes this shock is referred to as a "tie down shock". This means that at whichever corner this shock is attached, it has a delayed rebound effect, holding that corner down longer. That means this particular wheel will maintain traction for a longer period of time.

We will occasionally use a tie-down shock on the left front to help the car resist roll to the right rear.

Shock Dyno Testing

Not all shocks are made the same. There are manufacturing tolerances involved when building and assembling the precise parts that go into shocks. Chances are that if you pull four of the same part number of any brand of shock off the shelf, and dyno test them, you will find small variations in the compression and rebound curves between each. This is normal. You will also find that most shocks labeled as a 50/50 ratio actually having more damping force in rebound. This could present a problem when you are looking for a particular chassis tuning effect from a shock.

However there is a solution to this problem. We have each shock we use dyno tested by the manufacturer, and we keep those sheets with each shock. This enables us to refer to each dyno sheet to get exact damping velocity specifications, and it provides a base of comparison between different shocks. You know what you are putting on the car and can tune your car easier and much more effectively.

This has also helped with our working relationship with AFCO and lets us tell them more about what we are looking for and how each shock reacts in a particular situation.

Get dyno sheets from your shock manufacturer on each of your shocks (they will charge a small fee for this). Also, have your shocks dyno tested periodically because you can't tell if there may be an internal problem.

Chapter
10

The Braking System

There are several variables in a complete braking system. It starts with the pedal and goes to the rotors.

The Brake Pedal

Let's start with the brake pedal. The first consideration is the mounting position. We mount our pedals from a hanging position. This gets the master cylinders up higher than the calipers and therefore bleeding is simplified because the brake fluid runs consistently downhill. And, since the master cylinders are higher, they are mounted away from the heat of the header exit, and dirt and debris are not constantly getting on the master cylinders (assuming they are located on the outside of the firewall).

The pedal should include a balance bar to allow the driver to proportion braking force between the front and rear brakes. Pedals generally come in two different ratios — 6 to 1 or 7 to 1. A higher ratio (7 to 1) reduces pedal effort by the driver, but increases pedal travel. A lower ratio increases effort but reduces travel. We use the 6 to 1 ratio pedal with a hanging mount. The 7 to 1 is requested by some drivers who feel it is more responsive to foot pressure.

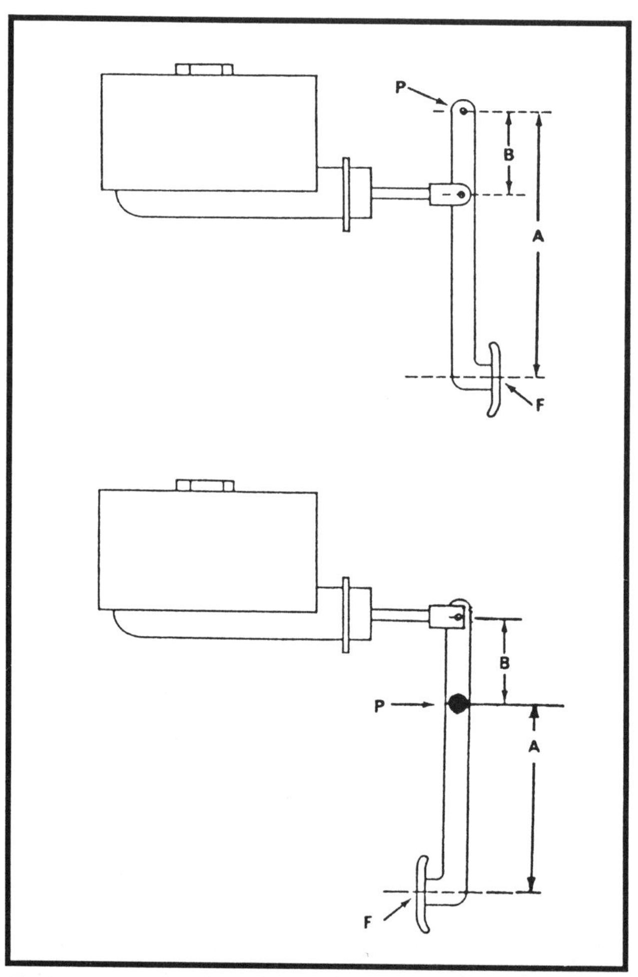

Pedal ratio is "A" divided by "B". "P" in the drawings is the pivot point of the lever arm, and "F" is the input force. Pedal ratio is a method of gaining more brake pressure with mechanical advantage by multiplying the driver's input force. As the pedal ratio increases, pedal pressure can be reduced, but pedal stroke increases.

Another pedal choice would be the reverse mount master cylinder inboard pedal. This particular pedal mounts the master cylinders inside the car. The advantage is that there will be no mud or debris getting to the pedal and master cylinders from outside the car. Everything can be kept cleaner. The only disadvantage to this system is the possibility of fluid leaking or spilling and getting on the floor or the driver's feet. As a safety measure, many racing asso-

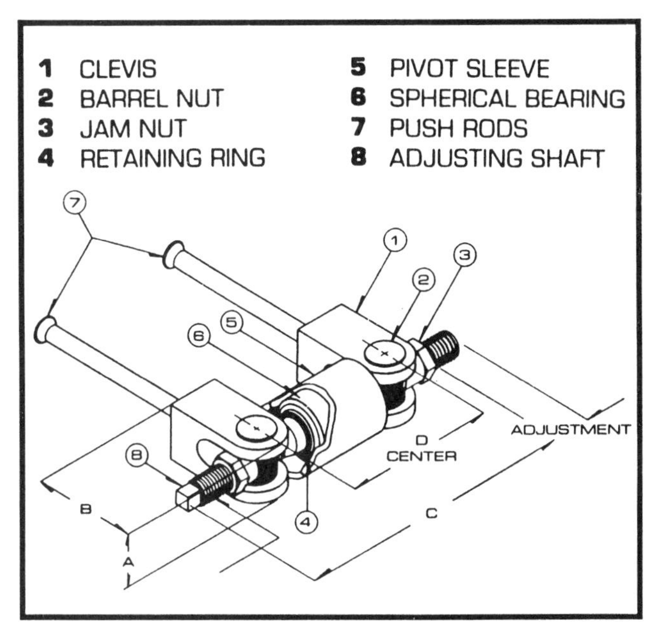

This is the balance bar assembly available from Tilton Engineering.

ciations require that master cylinders be mounted on the outside of the firewall (the engine side).

Once the pedal choice has been made, make sure the balance bar is working properly and the remote adjuster is mounted where the driver can reach it for easy adjustment. Make sure there is a solid mounting surface for the pedal mount so that the driver's braking effort does not flex the assembly. Also make sure that sheetmetal panels or other components do not interfere with the pedals and balance bar through the full range of travel.

Master Cylinders

The most important consideration is master cylinder piston size. We use 1-inch bore master cylinders in our system for both the front and rear brakes. They produce good smooth brake response. When we have used smaller cylinder sizes, such as .75-inch or .875-inch, the brakes become too responsive and have a jerky feeling. The driver will definitely have a better pedal response with the smaller bore and at the same time become a smoother driver. Some racers like to split the sizes using the .875-inch for the rear and the 1-inch for the front. This helps apply more pressure to the rear brakes, but when you have a balance bar that works properly, you can adjust the braking force distribution to the rear brakes or front brakes with a turn of the crank.

Brake lines should follow the shortest route from the master cylinder to the corners of the car, and should be neatly formed to the interior and frame.

Smaller piston sizes produce larger master cylinder output pressure, whereas larger piston sizes produce smaller output pressure.

There are all kinds of possible combinations with master cylinder sizes and caliper piston sizes. Each individual car builder or racer has their opinions on the braking system specs, but what we run works, and works well on our cars.

One other option on master cylinders is the remote type which has a remote reservoir located elsewhere on the chassis. These master cylinders can be put in a tighter area where close tolerances and clearance are needed.

Plumbing The System

Running brake lines is kind of like an art form. There is nothing as neat and trick as properly formed brake lines. The brake lines should follow the shortest route from the master cylinder to the corners of the car, and should be neatly formed to the interior and frame pieces. All of the lines should be attached with rubber-lined Adel clamps (see illustration). This helps insulate the brake lines from vibration.

For the hard lines, use standard automotive double wall steel tubing with 3/16-inch outside diameter. I have seen some cars with flex lines run all the way through the system, but this creates too much flex

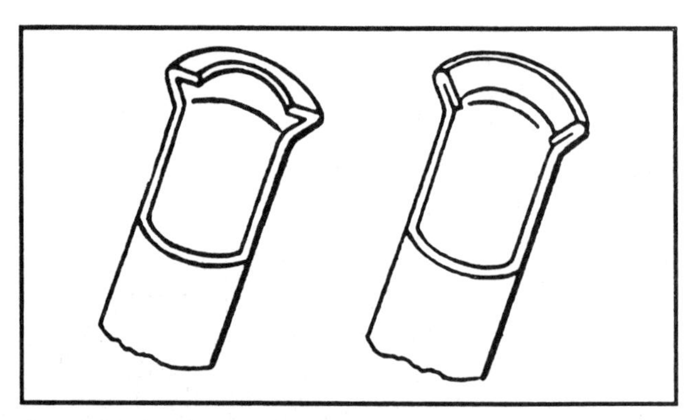

Steel brake line tubing should be double flared to prevent leaks under pressure. At left is the first operation of the double flare. At right is the finished flare.

The rubber-lined Adel clip protects brake lines, and electrical wiring too. The cushion assures tight fit and eliminates vibration.

and expansion in the system. The shorter the steel braided flex line, the better.

Make sure when bending the brake line tubing that there are no sharp bends or kinks. You don't want to restrict the flow of the brake fluid. Use a brake line tubing bender (available at parts stores or tool stores). Make sure all of the lines are routed away from heat, and are high enough to protect them from debris thrown up from the race track.

Make sure you slide the threaded male inverted flare fitting on the line before you start flaring. If you forget, you have to start all over again! Always use a double flaring tool for flaring. This system puts a double edge on the flare instead of a single. This eliminates any problems with cracking or leaks around the flare. Snap-on Tools makes a nice flaring tool kit if you do not have one.

Brake lines should be secured to the chassis at the point where they meet the flex lines. We weld a small metal tab at each corner of the chassis that mounts a fitting to the brake flex line. Then the 3/16-inch hard line screws into the fitting. This solidly secures the brake line at the point where the flex line meets the hard line.

Use stainless steel braided Teflon #3 or #4 brake flex lines to the calipers. This type of line is required to provide enough stiffness to prevent brake line swelling under heavy pressure. Be sure the flex line length is long enough in the front to accommodate full steering movement in both directions and full suspension travel. Be sure there is no interference with chassis components or tires. At the rear, also make sure the lines are long enough to provide a full range of suspension travel. Check to make sure the flex lines cannot crimp or kink or interfere with any other component through a full range of suspension travel. Make sure, when tightening the fittings, that there are no twists or bends in the flex line.

Brake lines should be secured to the chassis at the point where they meet the flex lines. Use stainless steel braided Teflon #3 or #4 brake flex lines to the calipers.

Be sure the flex line length is long enough to accommodate full steering movement in both directions and full suspension travel.

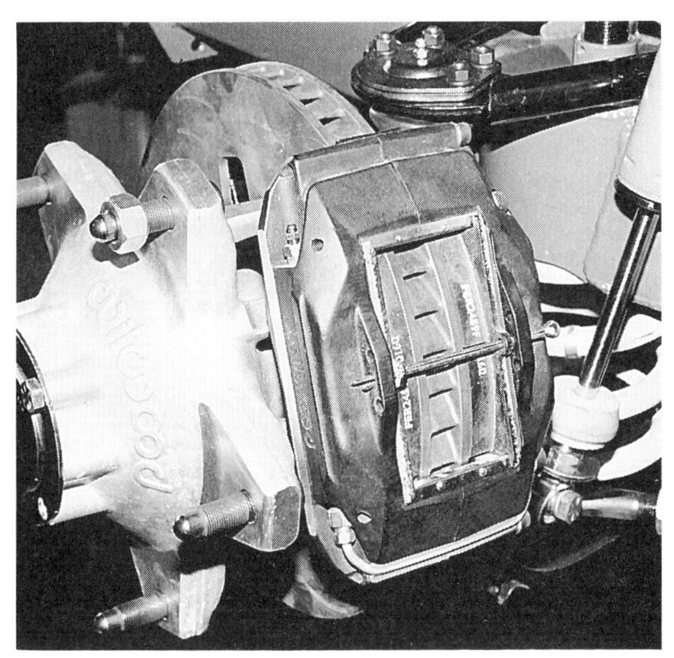

The Wilwood Dynalite II caliper.

Flex lines can be assembled with a swivel #4 female fitting on both ends, or a 1/8-inch pipe fitting on one end and the #4 swivel on the other end.

Calipers

Because braking is not as demanding on dirt tracks as on asphalt tracks, smaller, lighter weight calipers can be used. This would include the Wilwood Dynalite II, Sierra Mini GN Dual or Outlaw 2000. Each of these are a four piston design. They are all good calipers. We use the lightweight calipers to save weight even though the brake pads are smaller in size. There is plenty of stopping power here and the weight saving is substantial. There is no reason to carry around more unsprung weight than necessary.

We currently use the Outlaw billet calipers with 1.75-inch diameter pistons as standard equipment. It has a differential piston bore design which helps produce even pad wear. What this means is that the forward piston is a smaller diameter bore (1.62-inch) than the trailing piston so there is less pressure as the rotor enters the caliper and more pressure as it exits the caliper. At the point of rotor entry into the caliper, the drag on the brake pad is more extreme than when it exits. The smaller forward piston helps to equalize the wear. This works well because we have noticed equal pad wear versus the wear that we saw with the equal size pistons.

The Outlaw calipers have a differential piston bore design which helps produce even pad wear.

We also use a smaller diameter piston caliper on the right front. This lets the right front caliper generate less braking force than the left front, so the car automatically pulls to the left on corner entry (but not drastically). This helps the car enter and set up for the corner. Some drivers like equal front calipers because they want their car to be tight or extremely straight at corner entry. Our cars are naturally tight and that is why we use the smaller caliper on the right front.

A good braking system set-up would use four-piston calipers with 1.75-inch diameter pistons at the left front, left rear and right rear corners. Use a caliper with 1.38-inch pistons at the right front. This system works well on every dirt track condition.

Some of our drivers prefer to use Wilwood Superlite III calipers on the rear. These calipers are larger and weigh almost twice as much as the Dynalite II calipers, but they use larger pads and thus produce more stopping power. They prefer the Superlite calipers because they feel they lose rear brakes after a certain amount of laps. This will cure the problem, but I think they are relying on too much rear brake to turn the car. If you have the car set up free enough, you don't have to use as much brake.

Make sure you specify round piston seals for the calipers as they seal much better and do not give as much trouble as square O-rings. All caliper manufacturers install square piston seals unless you specify otherwise. Round seals let the pistons retract better. We were having problems with this using the

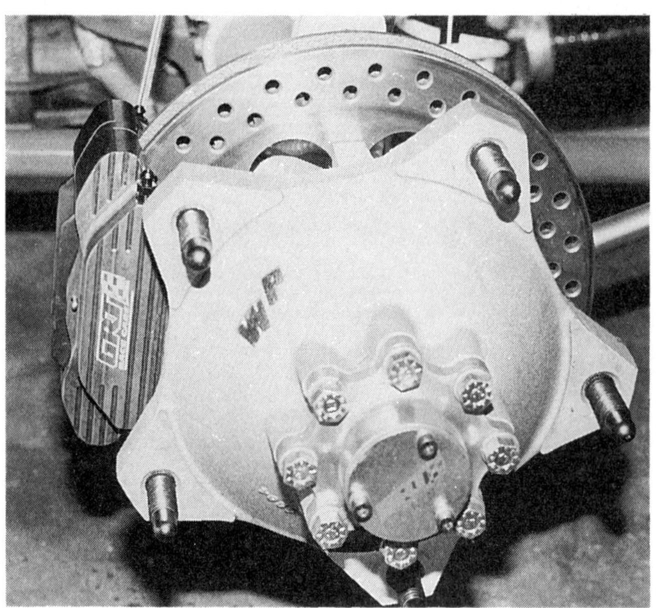

The rotor we use is the 11.75-inch diameter, 1.25-inch thick straight vane cast iron rotor, drilled all the way around in two rows.

square seals, and when we switched to round seals, our problems disappeared. This is very important for reducing brake drag.

Caliper Mounting Brackets

Brake caliper mounting brackets need to be solid. Thin brake brackets flex too much. Any time brake brackets are not stiff enough, they flex and the brakes cannot develop their full stopping power. A thinner bracket can also fatigue and crack, which will eventually cause failure and this can tear a lot of equipment up.

Brake brackets need to be made of at least .1875-inch thick steel, and preferably .25-inch. Our aluminum brake brackets that are on the bolt-on brake clamp brackets or brake floaters are .375-inch thick. If brackets do not meet these minimum specifications, they will develop fatigue problems.

Make sure when you locate the bracket for final installation that the caliper runs exactly parallel with the rotor. There needs to be an even gap between the rotor surface and the inner caliper edge. This is very important for even pad wear as well as maintaining the structural strength of the mounting bracket.

Make sure that the caliper is exactly centered over the rotor. The centerline of the caliper should be at the centerline of the rotor. This allows the pistons to push equally against the rotor when the brake is applied. This also prevents the rotor from rubbing the caliper. Make sure that all of the brake pad surface is being used on the rotor. You don't want part of the pads riding out in the air doing nothing. All of this must be done to ensure proper brake performance. This also increases pad life as well as stopping power.

One other area to check is that the pads are seated freely in the caliper. Be sure you can move them freely when there is no brake pressure applied. This ensures there is no pad drag when the brakes are not in use.

Check brake caliper mounting bolts periodically. For an extra margin of safety, you should safety wire these bolts.

Rotors

Rotors must run cool, should not wear excessively, and should provide excellent stopping power. They also must be light for maximum performance.

The rotor we use is the 8-bolt 11.75-inch diameter, 1.25-inch thick straight vane cast iron rotor. We drill the surface area of the rotor all the way around in two rows with a 5/16-inch drill bit. This lightens the rotor almost two pounds and also assists in cooling. The finished, drilled rotor weighs 7 pounds.

The other type of rotor we use is the .810-inch thick rotor, which is also drilled. This reduces weight even further, but being only .810-inch thick it does build more heat because the vanes and air gaps are smaller. This works fine for shorter races under 50 laps. But for longer races, the 1.25-inch thick rotor works best. When purchasing rotors I would recommend the 1.25-inch for almost every application.

Drilling Brake Rotors

We drill all of our own rotors and have not experienced any problems with them. We use a jigged pattern that we made to locate all of the holes to be drilled. Each hole is center punched after being located by the pattern, and then each hole is drilled with a drill press. If the pattern is not correct, the drill bit will miss the vane and will venture off to one side. The key is having an accurate pattern and a short, stubby 5/16-inch drill bit that will not flex like a longer one.

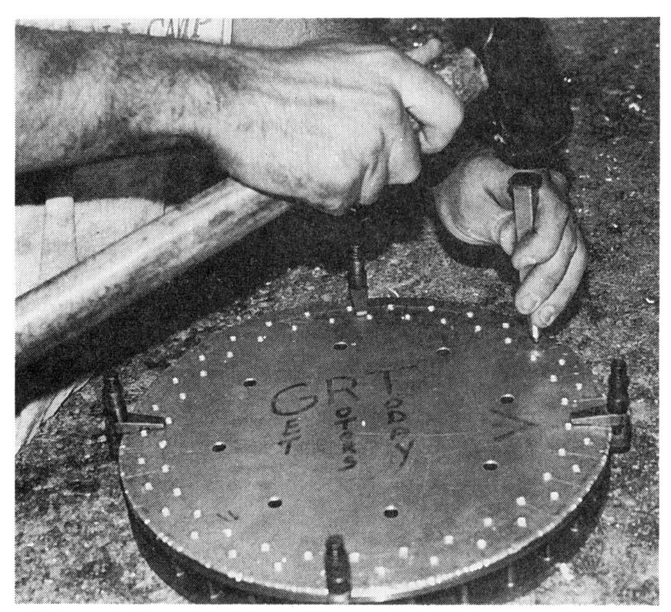

When drilling rotors, first each hole is center punched after being located by a pattern.

After the holes are drilled, the finishing touch is a chamfer of all the holes. The chamfering is very important because this keeps the rotor from wearing the pads. It removes the sharp edges left from the drilling. The chamfering also removes a small amount of weight. This also makes a difference in reliability because it prevents cracking, and pad wear is eliminated because the sharp edges are taken away. Rotors need to be checked periodically for cracking between the holes.

Carbon Carbon Rotors

Carbon carbon rotors, manufactured by Carbone Industries, are another option. Their major advantage is that they are very lightweight, but they are a very expensive set of brakes. They are available in .810-inch and 1.25-inch thicknesses, by 11.75-inch diameter. They require the use of larger, heavier calipers and special carbon carbon pads. The larger type of rotor is required in order to develop the greater clamping power required with carbon rotors and pads.

The system requires a duct system to help cool the rear rotors. Without the ducting, the rear rotors can get so hot they start to boil the brake fluid.

Maintenance is also increased. The carbon carbon rotors have to be replaced more often than cast iron rotors. And care has to be taken when washing the car. They don't like to get wet. They will absorb

A short 5/16-inch diameter drill bit is used to drill each hole.

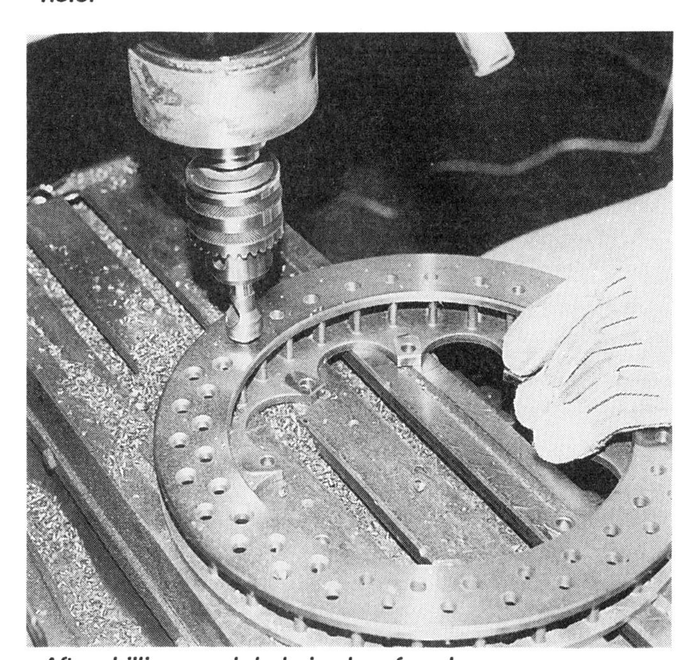

After drilling, each hole is chamfered.

water, so the rotors then have to reach a temperature past 212 degrees F. to boil it out.

Carbon carbon brakes require a warm-up period to obtain maximum stopping power. When we tested these brakes on our car, Bill Frye claimed he didn't have the brakes he needed early in the race or to qualify with. With all of the factors involved in running these brakes, comparing cost versus practicality we decided these brakes were not for us. The rotors are light and do work well, but we feel they are not a necessary cost efficient brake system. As a cost comparison, the carbon system costs $3000, as compared to a standard system cost of $400 (for four rotors and eight pads).

Brake Pads

What we look for in a pad is one that stops instantly and maintains stopping power throughout the race. You don't want a lot of fading. Also you must have a pad that wears well and is rotor friendly. Some of the pads we have tried in the past would eat rotors (cause excessive rotor wear).

We currently use the Performance Friction 83 compound pad. We have found that this pad outperforms any other pad we have tried. The 83 compound, which is semi-metallic, is rated for moderate duty stopping in a lower to medium temperature range. It will produce an average coefficient of friction of .48 through a temperature range of 400 to 1,000 degrees F. For a track that requires a heavy brake usage (and resulting high rotor temperatures), the 93 compound pad should be used.

Generally, pads should be changed when they are 75 percent worn. However, if you will be racing a 100 lap race on a heavy braking track, the pads should be replaced if they are 50 percent worn.

Disc brake pads should have a proper break-in process before they are used in a competitive event. Bedding-in new pads should be done by making several slower laps while applying the brakes frequently and gently, gradually increasing their usage with harder stops until their normal operating temperature is reached. Many times new pads will fade at this point. The brakes should then be allowed to cool for at least five minutes. This procedure properly heat treats the pad friction material. If it is not followed properly, and the brakes are used hard with new pads, the friction material will be "cooked" and the pads will lose their friction properties.

New rotors should be broken in using the same procedure as outlined above for bedding-in pads. Used pads should be used to break in new rotors. If new rotors are not broken in with the proper procedure, hard usage right away will cause thermal shock of the rotor which will reduce rotor life.

Many companies now sell brake pads which are pre-burnished or pre-bedded. This saves the racer the time required to bed-in new pads. With pre-bedded pads, go out and run a couple of laps and do some hard braking to let heat build up in the pads. This breaks in the new pads and also builds a transfer layer of pad material to the rotor, which is required for optimal braking efficiency.

When replacing pads, you will notice you have to push the master cylinder pistons in to get the new pads in. Always pump up brakes after this is done to make sure brakes are working properly. Also make sure the cotter pin used for pad retention on some calipers is replaced and not in a bind. On GM type brakes, make sure that the calipers are free and the rotors turn freely after pad replacement.

Brake Fluid

The brake system must have a good fluid to make all of these components work properly. Racing brake systems have an extremely high operating temperature and therefore the fluid must be able to handle this environment. In selecting brake fluids, there are two critical elements to consider — the dry and wet boiling points of the fluid, and the viscosity of the fluid.

The United States Department of Transportation (DOT) sets minimum standards for wet and dry boiling points of brake fluid. Always use a fluid that at least conforms to the DOT 3 standard, and much preferably exceeds it. The dry boiling point means the brake fluid is pure with no moisture contamination. The wet boiling point of brake fluid is after it has been saturated with moisture. The DOT 3 requirement for wet boiling point is 284 degrees F., while the DOT 4 minimum point is 311 degrees F. The dry boiling point for DOT 3 fluid is 572 degrees F. in a non contaminated form. This is brand new brake fluid in a sealed container.

There are several racing brake fluids available that exceed these requirements. Both Wilwood Engineering and Sierra Racing Products offer a brake fluid with a minimum dry boiling point of 570 degrees F. In addition, the Wilwood fluid is a low viscosity formulation, which aids in brake bleeding and lowers the chances of fluid foaming due to excessive pedal pumping.

Glycol based brake fluid absorbs moisture from the atmosphere very quickly. This can significantly lower the minimum boiling point of brake fluid. This

Lever style proportioning valves are much easier for a driver to use than a twist type. The lever valve clicks into place so the driver can feel each position.

is why it is so important to keep the fluid fresh and try to keep moisture out of the system. The fluid in both the can and the master cylinder should be exposed to the atmosphere for the shortest possible time. Never re-use fluid after it has been bled from the system, even if it is new. Purchase brake fluid in small cans so the contents can be used up all at one time. Most racing brake fluid is sold in 12-ounce containers.

DOT 5 brake fluid is silicone based. It does not absorb moisture and has a higher dry and wet boiling point. However, brake part manufacturers do not recommend dot 5 in a race car application. It has a higher viscosity, which causes drag in the calipers. And, it is highly expansive under higher temperatures, which causes more compressibility, and that means a spongy brake pedal.

Always flush out the braking system periodically with new fluid. And, if you ever experience severe temperatures with the brakes, flush the system completely and put in fresh fluid.

When bleeding brakes for the first time, we always fill the system up and open the bleeders with the master cylinders higher up on the car than the calipers. It will bleed automatically. When you bleed the brakes, make sure that you don't ever let the master cylinders run out of fluid. This will allow air to enter the system.

Brake Proportioning Valves

Brake proportioning valves help to tune a braking system for certain conditions. If you want to decrease a certain amount of pressure to a caliper, the proportioning valve will do this in percentages. Most proportioning valves decrease pressure in a range of 10 percent through 60 percent.

We usually install a proportioning valve on the right front to help turn a car into the corner by restricting pressure to the caliper. You can tune this once the track gets dry and not have such a severe decrease in braking power as if it were completely shut off. Normally, as the track dries out, the driver will increase the percentage of brake force to the right front so the car will stay straight under braking at corner entry. We like to have a shut-off valve also installed in the same line so we have the option to shut the right front brake off completely when the track surface is wet . Remember that proportioning valves are a tuning device and not a cure for handling problems.

If you use a smaller right front caliper piston size, then you already have less braking force going to the right front. With this situation, a proportioning valve is not necessary for the right front. We normally just put a shut off valve in the line.

Brake Shut-Off Valves

A brake shut-off valve is used at the right front to prevent braking force from locking up the tire on a slick surface. There are two different types of shut-off valves available — a manual and an electric.

A manual shut-off valve is lever actuated. An electric valve uses a toggle switch to actuate a valve which prevents fluid flow to the right front brake line. We use a manual shut-off valve.

You have to be careful when using a shut-off valve because you can cause it to lock the pad to the caliper. Always keep your foot off of the brake when turning valves on or off so that the pad is in the retracted position. If you have your foot on the brake when you turn the valve on, it will stay locked because it will not let the valve relieve the pressure.

A leaking shut-off valve will also create a lock-up problem. Every time you apply brake pressure the leaking valve will start building up pressure. Therefore it will start locking up the right front brake and you must turn the valve on and off to free it up again.

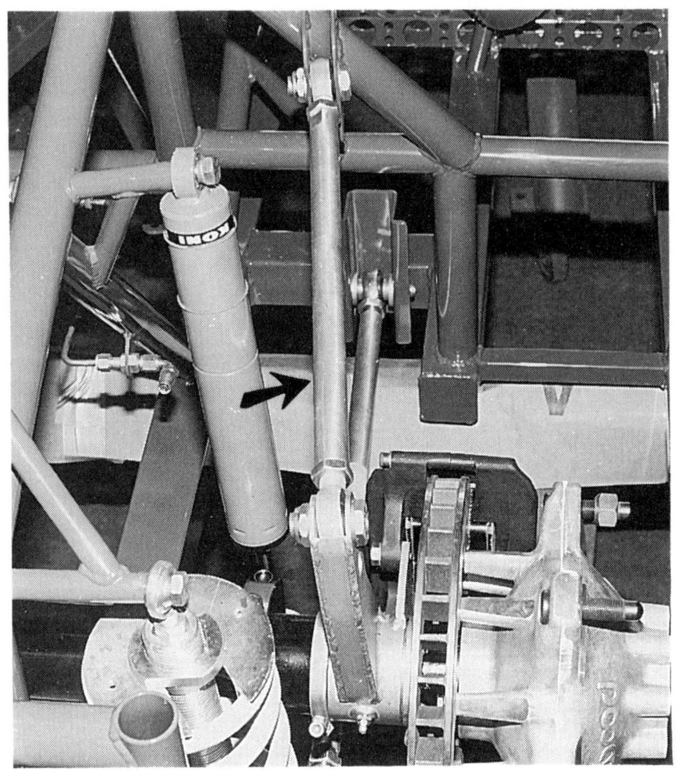

Brake floaters serve to remove braking torque forces away from other suspension movements and direct the braking forces straight into the chassis.

Brake Floaters

Brake floaters serve to remove braking torque forces away from other suspension movements and forces. They are separate brackets which are free to pivot about the rear axle housing, and are not welded to the axle tube. The brake caliper is bolted to the floater bracket, and a radius rod attaches the floater bracket to a chassis bracket. The radius rods direct the braking forces straight into the chassis and remove them from the suspension system. Adjusting the angle of the brake radius rods allows the braking forces to be tuned to individual track conditions.

The radius rod angle can be set with the front angled up, straight ahead, or angled down. Each different angle creates a different effect under braking on the tire at the corner of the car. For example, if the radius rod is angled up, the resultant force from braking adds force down on that tire contact patch. If the radius rod is angled down, the resultant force will lift up on the tire under braking. If the radius rod is set straight, there is no resultant force on the tire

Brake floaters are another tuning device that can be used to tighten or free up a car on corner entry braking. As you start raising the rod, when brake force is applied it starts to load that tire. The more angle, the more load. If you want the car to be free or turn on entry, then keep the right rod level and start raising the left rod until the desired effect is obtained at corner entry. For the opposite, raise the right rod up and keep the left side level to tighten the car under braking at corner entry. Adjusting theses rods is critical, so be sure you do not make big changes at one time. Learn what each adjustment does and remember that this too is a tuning tool, and not a cure for a bad chassis setup. We do not run brake floaters as it is another area that you have to work with and adjust.

Brake System Tips

Keeping the brake system maintained and fresh will provide the best performance and reliability from the brake system. Keep the moving parts lubricated, such as the balance bar, rod end bearings, and master cylinder pistons. Make sure the balance bar works freely. Check pads and pad wear. Keep a check on brake fluid and change it periodically. Bleed brakes often. Check all fittings for any signs of leaks. Check the rotors for cracking or heat checking. Make sure that the calipers release when you let off the brake. Brakes are a very important part of a race car, and if they are not right you will always have problems. Do not neglect the system.

Chapter
11

Tires and Wheels

Tire Selection

Tire selection is one of the most important factors in finishing the race car set-up and delivering it to the ground. Choosing the proper tire along with tire compound, tread design, grooving and siping patterns and sidewall stiffness are all factors to aid traction and performance.

Tire Compound Choice

You can choose from a variety of compounds from very soft to very hard. The proper tire hardness is determined by the track conditions. On dirt tracks, conditions can change greatly during the night from the beginning of hot laps to the feature race. Therefore sometimes you must choose a compound that will be effective at the end of the race. It may not be as fast early in the evening, but if you choose one that is too soft, you won't be in the ball park at the end.

If you question whether your compound choice will be too soft or too hard, always lean toward being a little too hard rather than too soft. A harder compound tire will be there at the end, whereas a softer one may give up.

In most cases, soft tires are used for qualifying, heat races, wet tracks, or tracks that don't have a lot of abrasion. Softer tires can also be used on harder tracks that don't build a lot of heat in the tires, or that have loose surface dirt. You will also notice that softer tires wear at a faster rate and will tear or rip easily. This is something you have to take into consideration when you use a soft tire. You have to learn how long the tire is going to do it's job on a particular race track condition.

The more you groove and/or sipe a tire, the faster a softer tire will start wearing. This is usually the reason that grooving and siping works well on a tire for qualifying and/or short heat races. It helps the tire get a bite. Tracks that have moisture in them and are packed well may not require as much tire grooving. If the track has loose dirt on it that is not packed down, grooving the tire more will increase traction.

Tracks that have a lot of rocks or pebbles promote wear at a more rapid pace, so take this into consideration when choosing a tire compound. This type of track is going to be hard or abrasive, so this will require a harder compound tire. A track that packs down and develops an asphalt type of surface will also require a harder compound.

A race track that is hard but yet doesn't get abrasive or build a lot of heat in the tires might require some tire grooving or siping. A track like this will sometimes require a hard compound tire. This in turn might cause a tire to slick over or glaze. When you groove a tire, it helps the tire dig into a hard surface and helps prevent glazing. Grooving and siping adds flexibility to the tread blocks, which helps prevent the glazing problem.

Watch track conditions closely. If you are not familiar with how a track might change, find out from people who race there regularly. When you choose your compound, whether it be hard, soft, or somewhere in between, if there is a doubt about which one to use, go with the harder of the possibilities.

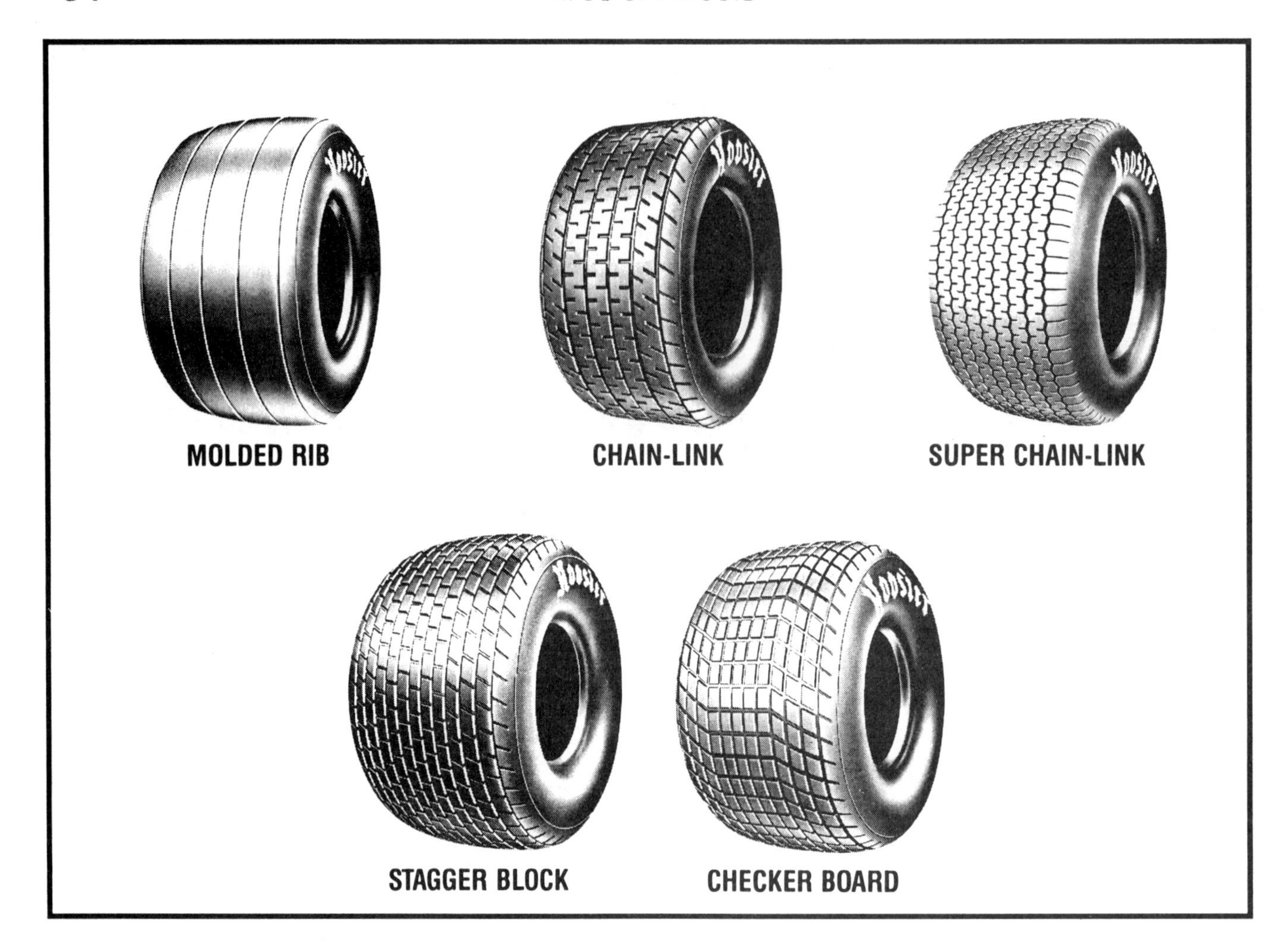

Tire Tread Patterns

Tread patterns on tires are usually formed in some sort of block or chain link configuration. These tread designs work well on average conditions. Other tire tread patterns that are used include a "rain tread" similar to a street tire, and a circle groove, which has multiple grooves around the circumference of the tires.

The tires with the block type of pattern need less grooving than a circle groove tire. But, in most conditions, all tires will need some kind of grooving or siping to promote traction. The only time you would not want any additional grooving would be on a hard asphalt type of track that is abrasive. On that type of track, you need as much rubber on the tire as possible to grip the track.

Tire Grooving & Siping

Once you understand the track conditions, you must decide how many extra grooves and sipes to put in a tire. Nobody really likes to groove and sipe tires, but it is a job that is required to make a car fast and competitive.

Before you start, make sure you have the proper equipment to do the job correctly. Good grooving irons are a must and make the job so much easier. To do the job correctly, you need an adjustable grooving iron which can be set to various temperatures. The groover we use is called a Rillfit IV and it is the "Cadillac" of all grooving irons. This particular grooving iron doesn't apply the heat until you actually push the groover through the tire. As you push the blades through the rubber, it cuts the groove while the fresh rubber ahead of the grooving iron keeps the blade cool enough so it does not get red hot and break a blade.

When working with a conventional groover, or setting blade depths, be extremely cautious not to burn yourself. It takes a long time to cool off after you unplug it, so don't lay one around where some-

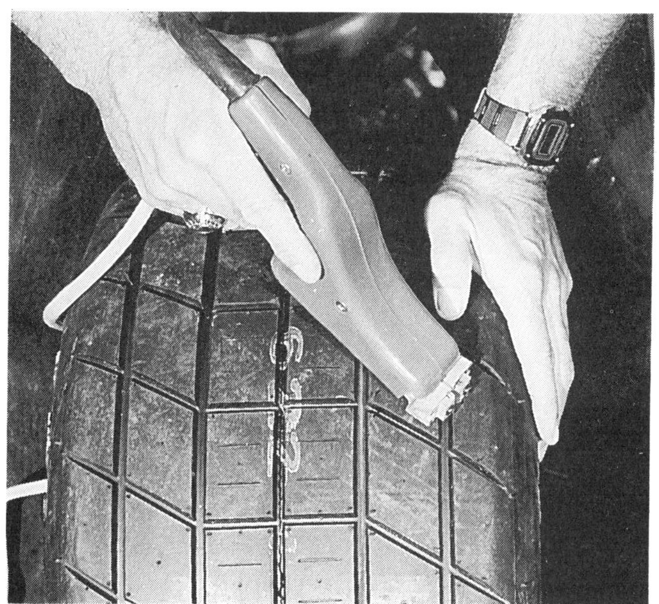
Good grooving irons make the job so much easier. Before starting, make sure the tire is clean and no dirt is on it.

one will accidently pick it up, or will burn whatever you lay it on.

Siping is done by turning the square cutting blades around backwards on a conventional groover, and letting just the sharp edge protrude through. Sipes are very thin cuts which are used to make a tread surface more pliable, to help prevent tire glazing, or to help dissipate heat out of tread blocks. Siping, when used with different conditions and patterns, lets the tire adhere to the track surface better and dissipates heat at a faster rate.

Before starting the grooving, make sure the tire is clean and no dirt is on it. Dirt dulls the blade and makes grooving difficult. You also need to spread talc powder across the area you are about to groove. This makes the blade slide easier across the tread pattern. Also, make sure the grooving iron is fully heated if you use a conventional grooving iron, or the heat range is set high enough on the deluxe model. If the heat range is not set properly, the groover tends to tear and stretch the rubber instead of cutting a good clean cut.

Different Types Of Grooves & Sipes

Grooving the tire's tread blocks puts more flexibility in them. This improves traction. The more traction that is required on a particular dirt surface, the more that a tire is grooved. Different types of groov-

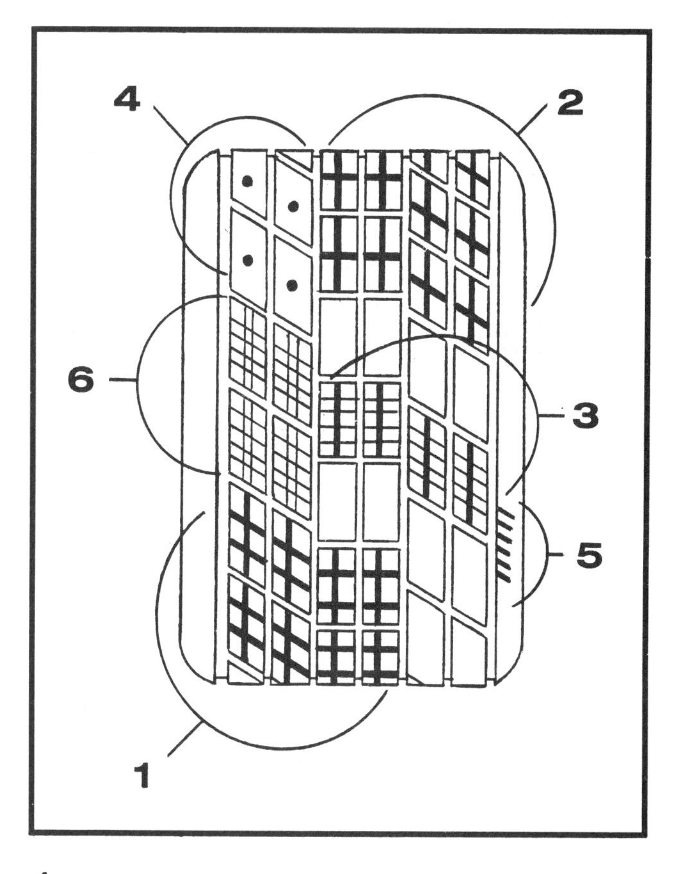

1 Vertical grooves and cross grooves are used on wet tacky tracks. They help to clean away loose dirt and let the edges of the blocks dig in for maximum traction.

2 On a tacky firm track that is packed down and not loose, use less grooving. One vertical and one cross cut in each block should be 3/16-inch wide.

3 On a hard track that doesn't build heat in the tires, use one 3/16-inch vertical groove with double cross pattern sipes. The sipes add traction and help build heat in the tires.

4 On a hard abrasive track surface, tires wear fast and heat builds up quickly. Don't modify the tread blocks oher than by adding a small hole in the middle of each block to let heat escape. The hole is 3/16-inch OD, drilled to 75 percent of the tread depth.

5 Grooving the sidewall on the outside helps a tire break up loose dirt on a track surface, and works well on a track with a cushion.

6 Use one vertical sipe cut and two cross sipe cuts for a track that is getting dry and slick, but does not have a shiny surface.

This is a molded rib or circle groove tire that has been grooved for use on the right rear on a track that is slick and abrasive. There are no sipe cuts in it because of a concern for getting too much heat in the tire or chunking the tire.

ing can tailor the tire to the track's surface. Here are some examples of different types of grooves and sipes to use on a tire for different track conditions:

Wet, tacky track early in the race program with loose dirt

Groove a lot of cross pattern grooves 1.5 to 2 inches apart. The grooves should be .25-inch wide. This will help clean away loose dirt and let the edges of the grooves dig in for maximum traction.

Tacky firm track that has packed down and is not loose

Use less cross grooves than with the loose tacky track pattern. Cut them 2 to 3 inches apart and make the grooves 3/16-inch wide.

A hard track that doesn't build heat in the tires

Use 3/16 to 1/4-inch wide grooves placed 1 to 1.5 inches apart with double cross pattern sipes.

Hard abrasive asphalt type track

Many times this type of track requires the use of circle groove tires with no side grooving. If the tire tread needs more flexibility to gain better traction, add some sipes in the tread about 3 inches to 6 inches apart.

Hard abrasive track that builds extreme heat.

Some types of track surfaces will increase tire heat, even to the point of blistering the tires. Grooving can be used to release built-up heat in the tire, cooling it down and preventing blistering. Grooves will divide larger tire tread blocks into smaller blocks which produces more air circulation and heat radiation around the blocks.

This tire has one fine cut down the middle of each block, then lots of cross and vertical sipe cuts. They are used to get the tire to build up heat.

The hard abrasive track surface also wears tires fast so in order to help the tire last, a minimal amount of grooving is used. You might just groove in the center of a block instead of going all the way across. Abrasive or rough tracks can easily tear the tread blocks on a tire when they heat up, so care has to be taken not to modify the tread blocks too much. Bill Frye uses a drill bit that drills a small hole in each block which lets heat escape and yet maintains maximum strength at the same time. A 3/16-inch drill bit is modified with a stop on it so the bit will not penetrate too deeply. The proper depth is 75 percent of the tread depth.

Other Grooving Tips

All of the grooving we have talked about here uses basic square-shaped grooves. V-shaped grooves work well on tracks where you need grooving for traction early, but as the tire wears, the groove gets smaller and more tire is on the ground for maximum traction. A V-groove would be used for a track that gets more hooked up and abrasive as the night goes on.

Another way to lengthen tire life is to not groove as deeply into a tire. In the early stage of a race, the grooves are available to help traction. As the tire wears, the grooves decrease and eventually disap-

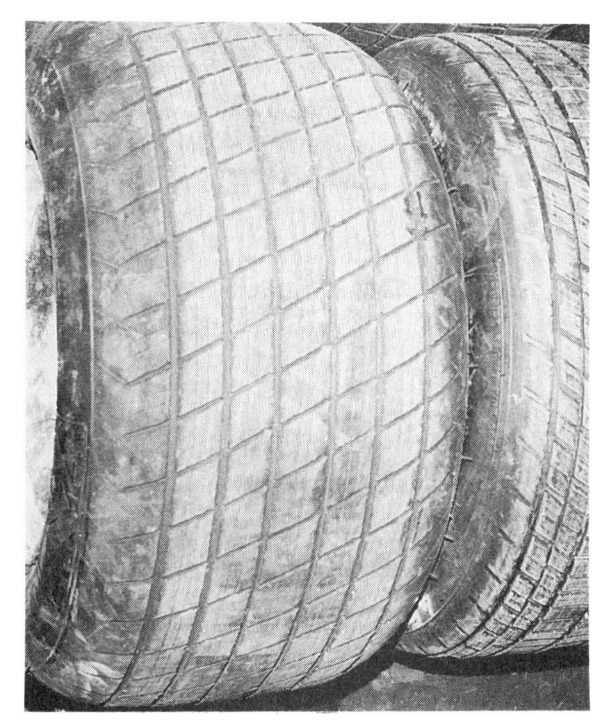

Grooves were cut in this molded rib tire for the right front to be parallel with the track when the car is back steered.

pear. This type of grooving is ideal for a track surface that requires a lot of tire surface in the late stages of a race.

Grooving the sidewall helps a tire break up loose dirt on a track surface, and works well on a track with a cushion. Run the groover straight across the shoulder edge of the tire at the same angle as the inside blocks. Be careful not to cut too far around the edge so that the sidewall is not damaged. The grooving depth would start out the same as on the tread face, but taper down to nothing as the shoulder edge is approached.

Proper Angles For Grooving

Grooving tires at angles instead of straight across depends on whether the car will be driven sideways more than straight on a race track. The back end of a race car will always be operating at some angle, depending on track conditions. A very heavy wet track will see the rear of the car hanging out a lot more than on a very slick, asphalt type of track.

The grooves which are cut into the tires must be placed at an angle which is consistent with the operating angle of the car. If a tire is grooved for a dry slick asphalt type of track, the grooves will be at

If you don't have time to run some laps on new tires, use a sander on them to scuff the shiny surface off.

much less of an angle than those which are cut for a heavy, wet track surface.

The reason for angled grooving is to keep the grooves perpendicular to the direction of travel. This allows the grooves to do their most efficient work on the track surface.

Front Tire Grooving Tips

Front tires perform better with a circumferencial groove. Since these tires steer the car, a lot of cross grooving is not necessary. The grooves around the circumference of the tire put some flexibility in the tread blocks, and this helps increase traction. The edges of the circumference grooves will bite into the track as the car is steered, improving the steering response.

Tire Break-In

Breaking in or scuffing tires is important for prolonging tire life and improving their traction capability. We believe that running a few laps on a tire and getting that smooth edge off of the tire will help the tire gain traction quicker. In 100 lap events, tires have to be properly cured to prevent blistering and wear. We run tires in hot laps and build heat in them, then we immediately dowse them with cold water. This cures the tires and helps them last longer in a long race.

If you don't have time to run some laps on new tires, another way to scuff tires is to use a sander on them. We use an air sander and lightly run an 80 grit disc over the surface area of the tire to knock off the shiny film. We believe that scuffing a tire will be an advantage to your starts.

Tire Pressure

Tire air pressure on most late model dirt tires is very close on all applications, regardless of the type of tire used. I have talked to a lot of different race teams and most everyone will be within 1 or 2 pounds of pressure in each tire at most race tracks.

This is what works best under almost any track condition:

LF 8 to 9 PSI	RF 10 to 11 PSI
LR 5 to 6 PSI	RR 9 to 10 PSI

These air pressures should be checked before each race, and before each chassis set-up. Accurate stagger cannot be checked without correct air pressure, so this is very important.

When setting air pressure, make sure that the tire is properly seated all the way around the wheel. We inflate each tire to 25 PSI to make sure it is seated properly, and then let the air back down to the desired pressure.

These suggested air pressures will keep a tire bead properly seated under race conditions. The only wheel that requires a beadlock is the right rear. This may be hard to believe, but we only use a bead lock on right rear tires. You can use bead locks on all four corners if you desire, but it is not necessary.

However, if a track is extremely rough and there is a cushion, then you may need bead locks on both right side tires. Mud plugs (plastic insert hub caps) may also be used to prevent mud from building up in the rim which would cause a severe shake from being out of balance.

Air pressures used in IMCA modifieds are different from those used in other dirt track tires. Tire pressures for IMCA modifieds are a delicate balance between the perfect amount for the correct contact patch, and adequate pressure to keep the tire seated on the wheel.

A normal range of tire pressures for IMCA modifieds on a dir track would be:

Reading the wear patterns of a tire can give meaningful feedback about the performance of the chassis and tires. The grain pattern on the inside edge of this right front indicates too much negative camber.

Wet track:

LF 8 to 10 PSI	RF 12 PSI
LR 8 to 10 PSI	RR 12 to 14 PSI

Dry track:

LF 6 to 8 PSI	RF 11 to 12 PSI
LR 6 to 8 PSI	RR 10 to 11 PSI

Reading Tire Surfaces

You can tell a lot by how a tire surface looks. This will help you determine if the compound is correct and the grooving and siping patterns are correct.

If the rubber on the surface is peeling back and ripping away, you probably have too soft of a tire. If the tire has just started working on the edges and they look like they are going to start peeling back soon, then the compound is close, but should probably be a little harder with some grooving and siping added.

If the surface of a tire is glassy or shiny smooth, then the compound is too hard. If that is the case, and you are limited to using this particular tire compound, then start experimenting with grooving and siping. When you get to the point that the rubber is actually tearing or chunks are coming out of the tread, too much grooving and siping has been used.

For the optimum tread surface appearance, you are looking for a surface that kind of "grains over" and resembles a sandpaper type of surface. The rubber is working and forming to the track surface, but it is just hard enough to keep from tearing or peeling.

If the tire starts blistering or "bubbling," then you must try to get rid of some heat and start thinking about saving your tires with chassis set-up and driving style. This is the only option you have when you are running as hard of a tire as you can get and it still blisters and bubbles.

Tire Stagger

Tire stagger is dependent on the track conditions and the way the car is set up to work. We usually set our front tires up with .5-inch to 1-inch of stagger. This is pretty general throughout all classes of dirt racing.

Rear tires are much more critical. Stagger plays a big role in the handling of a race car, and you will have to learn when and where to use the proper amount of stagger. GRT cars like a lot of stagger in the rear. We use 5 to 7 inches of stagger at most tracks. Our cars are tight and they need the stagger to help the car turn and go through the apex of the corner. This is really critical when the track has any bite to it or when a track gets hooked up.

A lot of this has to do with driving style also. When you reduce the stagger, it will tighten the car. If the car already is tight, then the driver will have to compensate to make the car turn. This in turn causes the car to break loose or tail out. This is what you don't want. Using less stagger makes this problem worse. So unless the car is extremely free and you can drive a car with less stagger, the corner speeds are usually slower.

There are race tracks which we run that sometimes allow us to use 3 to 4 inches of stagger, but not often. Other cars might require less stagger. However, as a general rule, if the stagger is in the 4 to 6-inch range, you will be in the ballpark. Experiment with 1 to 2 more inches of stagger, and see how the chassis reacts with it on different track conditions. Keep good records of the stagger used at each track, and always check it before the car is scaled.

Pressure Relief Valves

Most dirt late models don't use pressure relief valves. We have tried to run them in the past and did not have success with them. We experienced a lot of inconsistent tire pressure with the valves installed.

The relief valves are designed to keep a tire at a certain pressure. When tires gain heat, they will maintain a pre-set pressure by means of the valve releasing excess pressure. The valve will shut off once the original setting is achieved. If you choose to use these valves, be cautious. Make sure they work properly and do what you want them to do.

Tire Inner Tubes

Tubes have about phased out of short track racing. Back in the early 80's we tried everything to keep our tires from going flat, including tubes. They are heavy, bulky, and hard to work with. The quality of today's tires and wheels have about eliminated the need for tubes. We suggest you go tubeless if at all possible.

Wheels

This is a subject that every racer talks about, including which style, what offset, what kind, beadlocks or no beadlocks. The first decision is what kind of wheel to run. Mostly this depends on the class rules for your application.

Steel Wheels

If you have to run steel wheels, then we suggest a quality lightweight wheel. The lightness of the wheel plays a major role in performance due to rotating weight. And, if wheels are lighter, they won't be as strong and therefore will fold up in a crash more easily.

This has it's advantages and disadvantages. If the wheel folds, it will not damage as many suspension components as it would if it were heavier. This is good. But, there is a fine line between going too light and not being able to finish a race because a wheel tore up and leaked air out, resulting in a flat. So when you choose a steel wheel, choose one that is both lightweight and durable.

There are a lot of good wheel manufacturers, and we suggest you check with your local wheel supplier

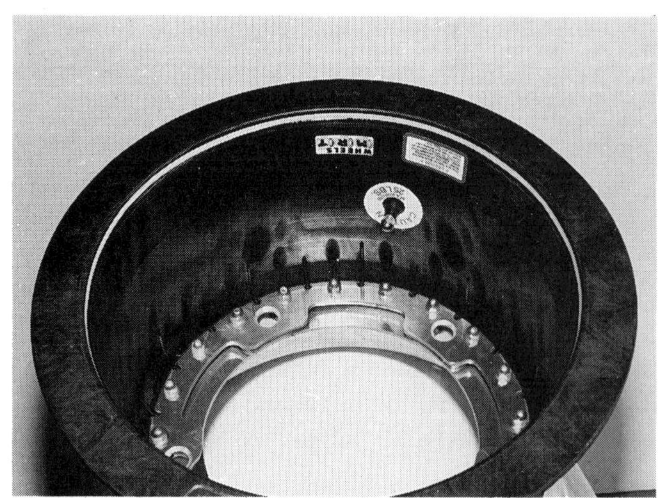

(Above and right) Carbon fiber wheels have an aluminum inner half and a carbon fiber outer half. They are bolted together in the middle with a series of small bolts.

about costs and warranties. Steel wheels are mostly used in sportsman and IMCA classes.

Aluminum Wheels

Aluminum wheels probably are the most common wheel used in late model dirt racing. We have used several different brands and sell two or three. We currently use the Weld wheel.

Aluminum wheels are light and durable. They absorb crash damage well and keep on racing unless it is a severe accident. One word of caution when shopping for aluminum wheel is to always make sure that you know what you are getting. Aluminum wheel manufacturers offer different thicknesses of material in their wheels, and some are truer than others. The thinner wheels are going to be the lightest and will cushion an impact better. So make sure to get all of the specifications on the wheels before buying.

Aluminum wheels come in a variety of widths, offsets and bolt patterns. In the wide-5 style wheel, the Weld XI series wheel has multiple bolt pattern holes that are used as lightening holes. This also assists in faster tire changes by not having to line up only one set of bolt holes.

Aluminum wheels also have a variety of beadlock styles and mud cap combinations. The current wheel we use has a ring that fits in between the beadlock ring and the wheel edge that has Dzus springs installed. This system lets a crew member install or remove the mud cover quickly. It is very durable and reliable when running against some of the most severe cushions and muddy conditions that are sometimes encountered.

Carbon Fiber Wheels

Carbon fiber wheels are another type wheel used in late model racing. Most carbon fiber wheels used have an aluminum inner half and a carbon fiber outer half. They are bolted together in the middle with a series of small bolts. A bead of silicone is used all the way around the two halves on the outside of the wheel to assure there are no leaks. This is done so that the outer halfs can be repaired easily without having to replace the entire wheel. The carbon fiber outside half is extremely tough and will reduce suspension damage when involved in a side impact. This is because the carbon half will fold up and crack, absorbing most of the blow. This is a big advantage as far as saving suspension components. MRT offers a repair guarantee on their carbon fiber wheels. Contact MRT or a dealer for details.

You as the racer will have to make the decision on what type of wheel you want, and what advantages and disadvantages you want, and the cost level you can afford.

Wheel Balancing

Wheel balancing is not a common practice on short track cars. If you go to any race involving late models or modifieds, you will notice they break the old tire down and mount a new tire. Then they bolt her on and race. There are several reasons for this. One of the main reasons is that there is no time to balance the tires, and most tire vendors never bring balancing equipment to the race track.

The track conditions and the length of tracks play a role in this too. Tracks are fairly short. And, the tires and wheels we use are true and round enough to not have to worry about balancing. Only on certain occasions have we ever balanced a tire. This is one area that I would not have much concern over.

Wheel Trueness

Wheel trueness is probably more important than any other part of the wheel's rotating performance (other than weight). All newly purchased wheels should be checked for lateral and radial run-out. Lateral run-out is the sideways movement, commonly referred to as wobble. Radial run-out is the true roundness of the wheel.

We check our wheels once or twice a year for trueness. What we do is bolt the wheel securely to a hub mounted on the car, and rotate the wheel using a dial indicator mounted on a jack stand or something equivalent. When the wheel rotates, take a reading from the wheel edge. We have seen wheels with as little as .030-inch run-out, and some with as much as .070-inch. The ones closer to .030-inch are the choice wheels. Each Winter when racing season is over, check your wheels. Replace or repair any wheels that are out more than when they were new.

Wheel Offsets

Wheel offsets commonly are available between 2 inches and 7 inches. At GRT, we have designed our cars using a 5-inch offset wheel. This makes life much easier for the racer and crew. You don't have to wonder about which tire goes on which wheel because of different offsets. All the wheels can be the same.

Obviously different offsets are required to tune a chassis for different conditions. To accomplish this, we use a wheel spacer to change the offset of our

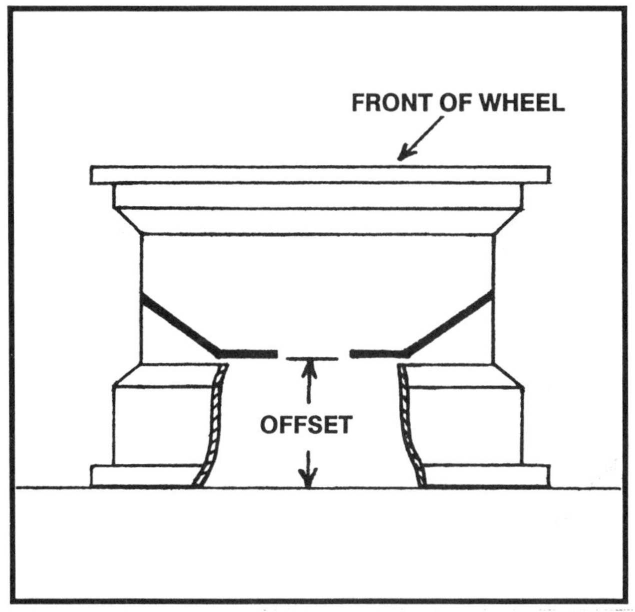

Wheel offset is the back spacing of the wheel. To measure, lay the wheel with the back side down on the shop floor. Measure from the floor to the back side of the center section of the wheel. This is the offset.

wheels. This really simplifies the process and does the same thing as installing a different offset wheel, provided the spacer is the same width as the offset you are changing. You can put a 2-inch wheel spacer on and bolt up a 5-inch offset wheel and have accomplished the same thing as using a 3-inch offset wheel. If you use a 1-inch spacer, you would have a 4-inch offset wheel. This is much easier than changing tires around on wheels.

No matter what kind of car you race, try to standardize the wheel offset. It will save you money and a lot of time.

When experimenting with offsets, make sure that the back spacing does not interfere with any suspension or brake parts on the car. Sometimes a back space that is too deep will cause a tire to rub or a wheel to hit a steering rod end, a shock or a spring mount. Make sure you check all clearances.

Also, changing offsets on wheels changes weight distribution, and changes the scrub radius on front steering. Take all of this into consideration when determining the offset of the wheels you choose.

Valve Stems

Most all of the wheel manufacturers install valve stems in their wheels at the factory. We suggest using

Bead locks will hold the tire on the rim with low air pressure.

the chrome shoulder type valve stem with a metal screw-on top. This top goes over the entire valve stem down to its base. This protects the valve stem from minor impacts or sharp objects hitting the stem. The cap has a small rubber gasket inside it that seals against the top of the valve stem. If air leaks past the valve stem core, this cap and seal will usually prevent the air from escaping.

To check for valve stem leaks, after the tire pressures have been set, we put a little soapy water on the top of each valve stem with a spray bottle. If there is a leak, it will blow out the soapy water. This practice ensures you don't have a leak in the valve stem areas.

Always keep extra valve stems and valve stem cores handy somewhere in your tool box in case you should experience damage to one.

Bead Locks

Bead locks are a life saver. They will hold the tire on the rim with low air pressure, and also serve as a mounting point for a mud cap. Bead locks can be used on any type of wheel. As long as the rules allow it, they should always be used on the right rear, and sometimes on the right front.

When mounting a tire, tighten the bead lock evenly and make sure the tire doesn't jump out of the beaded area. The tire has moved if the bead lock ring will not tighten evenly. When mounting bead-locks we always use a 3/8-inch drive impact wrench set at 25 to 30 foot/pounds of torque. After you have tightened them with the impact, check the torque with a hand wrench for insurance.

Sheet Metal Screws

Before the tire bead lock was invented, sheetmetal screws drilled through the wheel edge into the tire bead were used to keep the tire on the wheel. Using sheetmetal screws today is old technology. But, some racing associations forbid the use of bead locks. In that case, sheetmetal screws are the only solution.

The only type of wheel that you would want to use with screws would be steel wheels. You have to be careful to make sure the screws go directly into the middle of the bead. You can do this by drilling the pilot holes accurately. This method is a time consuming and troublesome way to hold a tire on a wheel. We don't recommend it, but if it is your only choice, do the best you can. There are no guarantees, and you will tear up the rubber bead on the tire the more you put screws in.

Wheel Maintenance

Wheel maintenance is important. Wheels that are bent or cracked need to be repaired or replaced. You need to check each wheel closely three or four times a season to ensure there are no problems starting to occur. And, always check the car's wheels thoroughly for damage if you have been involved in a wreck — even a minor wheel bumping incident.

Always rotate your wheel supply and get rid of wheels that need replacing. If you keep using a bad wheel, sooner or later it will fail and could cause a crash. This tears up your equipment, and someone could get hurt.

Things to inspect a wheel for are lug hole wear, cracks or bends, trueness of the wheel, and worn or bent wheel edges.

If you have bead locks, make sure you occasionally lubricate the bead lock bolts and check the flat washers to make sure they are not rounded out. Replace any bead lock rings that are bent or cracked.

Only professionally trained persons should mount tires. Before inflation, the entire assembly of any multi-piece rim must be placed in a restraining device. Stand away from the assembly when inflating. Never exceed 25 PSI when seating a tire.

Chapter
12

The Engine

Which Size Of Engine To Use

The big question is what type of motor to use. A lot of variables go into this decision, but first you must determine where you are going to be racing and what class you will be racing in. The decision can be made easier if you only run on local tracks on a weekly basis rather than travel around a lot and go from track to track. Obviously the racer who runs one or two tracks and knows the track conditions has an easier decision on which engine to run.

Let's assume you have the choice of a small (typically 362 cubic inches) or a big motor (410 to 430 cubic inches). The big motor is more expensive, usually requires more maintenance and creates more difficulty in controlling the torque and horsepower. Big motors usually work best on extremely fast and hooked up race tracks. The big motors have a place and have an advantage only in those places. For racing at one or two local tracks, the small motor has been a dominant force. Let me explain why this is true.

Racing on the road 50 to 60 nights a year, you learn what works and what doesn't. The sanctioning bodies are allowing weight breaks and some allow the use of a larger spoiler with the small motor. S.U.P.R. (Southern United Professional Racing) uses what they call a spec motor, which is a 362-cubic inch engine with a spec head and intake. When using this engine, you get to weigh less and run the 12-inch spoiler. Believe me the heavier cars with bigger engines don't stand a chance.

The Hav-A-Tampa series has a similar rule. Their small motor rule is 362 cubic inches, and that enables a racer to run a lighter car with this displacement. Hav-A-Tampa late models have been dominated by the small engines.

The small engine has more advantages. They don't have a large amount of torque or horsepower that makes the car harder to control. The driver can be smoother and more consistent. This really starts showing gains when the track gets extremely dry.

One of the largest complaints heard in dirt racing today is, "I can't get hooked up" or "no forward bite". That may be true to a certain extent, but these racers are using a 430-cubic inch engine with massive amounts of torque and horsepower. That encourages tire spin!

Even if your chassis is working properly, a car with a large engine is hard to control and the best drivers will tell you smoothness wins races. A smaller cubic inch engine would be better if you can only build one engine. Cubic inches and torque usually go hand in hand. The bigger the cubic inch number, the more torque it produces. Camshafts and compression affect when and where in an RPM range the torque is delivered, but generally more cubic inches produce more torque.

In a higher weight class, torque would be a benefit. The heavier the car, the more torque required to move the car. When you are building your engine, try to engineer a smooth torque range. Stay away from snappy torque curves.

Carburetion

Once you have developed a good smooth horsepower and torque range, you can start tuning your engine with carburetion adjustments. Carburetor size and jetting is very important to the overall

Always use a screen in front of the radiator to keep it clean and protect it.

performance of an engine. Most engine builders will be familiar with what size and jetting ranges are required for your particular engine. Make sure your fuel pressure is set correctly and the floats are adjusted properly.

One thing that we have always done on our cars is mount the fuel regulator so it is somewhere close enough to the driver's reach so he can adjust it. Install a #8 hose fuel return line in the car. Splice the line somewhere near the driver's compartment to mount the regulator there. Let the adjusting knob come through the interior panel so it can be adjusted by the driver. The advantage is that the driver can adjust fuel pressure at all times without assistance from a crew member. This allows an adjustment to be made under racing conditions. This could help if the engine is running hot and you need to richen the carburetor. It also can be leaned down for maximum performance.

Make sure the timing and valve lash is checked before each night of racing. We recommend that you check valve spring pressure once a week also.

Headers

Headers provide the finishing touches to an engine. I talked with Russell Baker of Russell Baker Racing Engines in Miami, Oklahoma. GRT and Bill Frye have been using his engines exclusively for many years. Through all of his years of research and development, along with our continual on-track testing, we have a vast amount of experience with headers.

The smaller tube and or smaller collector header produces more of a low range power band. This is more low RPM torque. If you have an engine that is a little weak on the low end, and you have a big tube header, you might try using a smaller tube.

With a large cubic inch engine, you will have a fair amount of torque. You will need a smoother, broader upper end. The big tube header (1.875-inch OD to 2-inch OD) primaries with a 3.5-inch collector will produce a much higher RPM power band.

Step headers, which switch primary tubing from one size to another, kills the low end torque and adds to the upper end more than just a straight tube header does. Currently most of the headers that Baker uses are straight tube headers.

Smaller cubic inch motors which need more low end torque or a good low-to-mid range power band will perform better with a smaller tube header (1.75-inch OD). As your engine selection starts to grow in cubic inches, you can start moving your torque and power bands to a higher RPM range with a larger diameter tube (1.875-inch OD to 2-inch OD).

Headers which are closer to equal length primaries will create a more efficient engine and a smoother power band. Ask your header manufacturer about how close to equal length tubes are on the headers you may purchase.

Zoomie Headers

The old style "zoomie" header, which has just a tube for each port then goes straight down and dumps the exhaust at the ground, has more bottom end torque than any other header. If you need this kind of torque, check with your local track or organization to determine if the rules will allow a non collector header.

Header Coating

You may want to consider getting your headers ceramic coated from a company like Jet-Hot. They use a ceramic/metallic coating on the headers inside and out. This makes a much more efficient header, improves header life, and prevents much of the exhaust heat from radiating out of the header pipes. The inner coating aids exhaust flow and prevents build-up of rust and carbon. The appearance is even improved due to the shiny aluminum look. The expense is greater (about $200 per set extra for the coating) but well worth it in the long run.

Engine Mounting

Mounting the engine to the chassis has been done in several styles. Usually some type of rear engine motor plate that mounts between the bellhousing or transmission and the engine block serves as the main support for the engine. Some plate styles are split and mount to one particular point on each side of the engine. We prefer the solid one-piece rear engine plates.

Recently we designed a new rear engine plate that actually ties in the frame area at the rear of the engine because of it's triangulated shape. This allows the engine plate to serve two purposes. It mounts the engine, and structurally strengthens the chassis. Our standard engine plate (like most others) just went straight across and tied into the chassis at a weak area. With the new engine plate, it spreads out to the top of the frame snout bar and down to the lower main frame, creating a triangulated effect which is extremely strong. The rear motor plate is fabricated from .25-inch thick aluminum.

Front engine mounts are normally the "L" shaped side block mounts. These front mounts need to be made of steel in order to provide adequate strength and rigidity.

Some racers use a front engine mounting plate which mounts to the front of the block. This usually interferes with a lot of components and requires a little more engineering to make everything mount properly. Front engine plates are not necessary.

If you are concerned about the front of the engine not being tied into the chassis using only side block mounts, you can fabricate a support brace that ties in the front of the head to the front bay brace with a piece of .375-inch aluminum bar. This idea was designed by Bill Frye and he uses it on all of his cars. This bar goes from the right front cylinder head (using one of the mounting holes already threaded in the head) to the shock tower bay brace. This brace just ensures there will be no movement in the front engine area. It ties everything together.

Transmission Mount

One other area of concern is the transmission mount. This ties the engine and transmission to the chassis for more support. It also reduces the chances of tailshaft flex which stabilizes the drive line. When mounting the transmission to a solid mount, make sure there is no force or misalignment on the transmission or mount when you bolt it together to avoid cracking the transmission housing. Make sure it fits precisely before you bolt it up.

Engine Location

Engine location — height and left offset — is another consideration when building a chassis. If you are not sure what you need, call a reputable chassis builder, or someone who has a good chassis, to find out. Our GRT chassis uses both a .625-inch to 1.625-inch offset to the left of the centerline of the chassis, and a crank centerline height of 8 inches above the bottom of the frame. The engine has the two lateral mounting positions built in to adjust for track conditions. The 0.625-inch offset position is standard and is used for most all applications. The 1.625-inch offset position is used for a track that has an asphalt-like surface and has a lot of rubber.

I feel an engine doesn't need to be mounted extremely far to the left to make the chassis roll more. However, mounting it further to the left works better on abrasive, hooked up race tracks.

Our engines set relatively higher than most cars. This is because it helps the car transfer weight left to right and front to back easier. Don't go to extremes, but some engine location experimenting could be beneficial.

We use the maximum allowable engine setback to keep excessive weight off the nose of the car. This also gains rear weight percentage, which helps forward traction. Most late model rules specify a 6-inch setback. That is measured six inches from the number one spark plug to the center of the upper ball joint. This is the best location we have found.

We have also found that you can get the front of a car too light, and this will create a push condition. We got the front lighter by moving the engine back further, and used an aluminum engine block to reduce the front weight. These modifications made the car perform worse because the front end was not sticking properly.

Chapter
13

Transmission, Drive Line & Rear End

The Bert and Brinn Transmissions

The Bert and Brinn transmissions have got to be the best thing that ever happened to dirt track racing. The operational characteristics of these two different transmissions are virtually identical. They are almost bullet proof. They are economical and lightweight, have an internal clutch, are direct drive when put in gear, and will not slip. They have a very low rotating weight, and weigh only 49 pounds.

These transmissions have no mechanical actuating linkage, but have a hydraulic cylinder that engages the clutch. These transmissions use a motorcycle type of clutch that, when engaged, drives the car forward or backward depending on which gear you have selected. It works this way only in the low/reverse side. In other words, when the driver pushes the clutch pedal in, it starts engaging the clutch and making the car move. While the clutch is fully engaged with the pedal all the way down, the car is moving. The hydraulic cylinder in the transmission has engaged the clutch discs and is turning the low/reverse gears. When the engine RPM synchronizes with the wheel speed, you let off of the clutch and shift into direct drive. This is done much in the manner as shifting a big truck without the clutch. Shifting moves a double-sided dog clutch, and the input shaft and output shaft are locked directly together. The power transfer does not flow through any gears. Once you get the car in gear, the only way to stop the car is by pulling the transmission out of gear. In other words, the car only has a neutral position when it is not in direct drive. The transmission is also in neutral while it is in low or high until the clutch is depressed.

The tailshaft mount on a Brinn transmission uses two bushings made from .75-inch OD, .120-inch wall tubing. When these spacers are installed, make sure they don't put any force on the tailshaft housing. The distortion could crack it. The spacer length has to be precise.

I've seen these transmissions used for five or six years without failure. The Brinn transmission uses the same output spline as the Muncie and Borg Warner transmission, and the same driveshaft length as a Muncie four speed. The Bert, however, requires a 2.5-inch longer driveshaft. In our opinion, if your class will let you run a Bert or Brinn transmission and you don't have one, you are at a disadvantage.

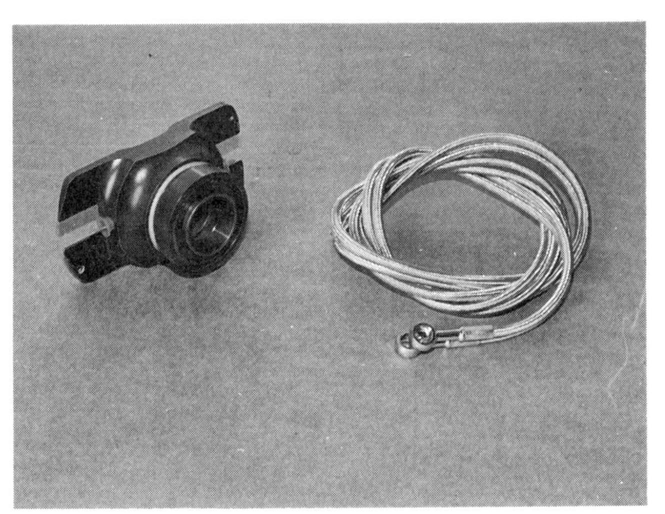

Using a hydraulic throwout bearing, such as this one from Tilton, eliminates having a slave cylinder and a clutch fork.

Other Types Of Transmissions

There are other transmission options if you do not run a Bert or Brinn transmission. In all the years I've been in racing, I've seen every type used, and the Muncie four speed is the best transmission to use. It should be used in conjunction with a quality triple disc clutch. The Muncie transmission should be lightened and modified by taking first and second gear out of them. Use a hydraulic slave cylinder to engage the throwout bearing instead of a mechanical linkage. The trick here is to get the clearances correct on all of the clutch discs and the throwout bearing-to-clutch diaphragm. You don't want anything dragging or not engaging fully.

For IMCA and Sportsman or any other classes that require a stock type transmission, a good aluminum-cased Muncie four speed that has been reworked is good, provided you have a good clutch to go along with it.

Using a hydraulic throwout bearing eliminates having a slave cylinder and a clutch fork. These are made by Ram, Tilton, Quarter Master and Howe. Not all bearings are compatible with all clutches, so check with the manufacturer about your application. Any time you reduce moving parts and linkages, my theory is it has got to be better.

Powerglide Automatic

Another option is the Powerglide-based automatic transmission. Most of the IMCA modifieds that we build use this transmission. Those transmissions have been developed to the point where they are relatively light and extremely reliable. A racing-prepared Powerglide weights about 95 pounds — much less than a standard transmission and clutch assembly. My IMCA house car driver (L.T. Davis) has run a TCI Powerglide transmission for years and he swears by them.

Some of the modifications used in the TCI competition Powerglide include optional low gear planetary sets (we use the 1.82 ratio), an external transmission cooler and a finned aluminum pan. An aftermarket steel clutch hub replaces the steel cast iron part. Using this part also allows the installation of extra clutch plates. Extra internal rotating weight is removed, and a high strength steel alloy input shaft is installed. A push start slave cylinder mounts on the right side of the transmission, which allows the car to be push-started.

The very low rotating mass helps a race car decelerate and accelerate much quicker. This means a driver can drive deeper into the corners, and start accelerating quicker.

The TCI valve bodies eliminate the need for external valving of line pressure when using a straight pump drive. This design allows you to operate the transmission by using the stock detent linkage with pedal operation. A clutch pedal is connected to the transmission detent. Depress the pedal, place the transmission in gear, ease out of the pedal, and the line pressure is released.

Proper mounting of the Powerglide is very critical. The rear of the engine should be mounted with an engine plate, and the rear of the Powerglide should be supported by a rubber mount at the rear of the case.

Shifting

Shifting all of these types of transmissions is as important as having the right transmission. They make good quality shifters for each transmission. Let's talk about the Bert or Brinn first. They require a basic two lever shifter that has two shifting rods. One engages low/reverse, the other high/direct drive. These are very simple, and there is also a model with a lock built onto the shifter. This acts as a safety to make sure the transmission stays in gear in case the transmission has a failure and jumps out

The Bert or Brinn transmission uses a basic two lever shifter. One engages low/reverse, the other high/direct drive. This model has a center lock built into the shifter.

of gear. You can buy these shifters at any oval track supply store. This type of shifter also works well on Muncie and other standard transmissions when they have been modified and only have high, low and reverse.

There are three lever shifters that let you shift the three speed or four speed transmission. They are a bit confusing because you must pull the transmission out of the first or second gear to neutral position before you shift the third or fourth gear. It is just not necessary to have a three or four speed transmission on short tracks today. We recommend the two speed transmission with a low speed and a high for short track racing.

On the other hand, if you have an automatic like the Powerglide, there is an excellent single lever shifter that has stages on different slots in the shifter housing which assures you that you are in the right gear and that it will stay there. RTC (Racing Transmission Components) is a prime example of this type of shifter. We use their shifter exclusively in our IMCA cars.

Transmission Maintenance

No matter what type of transmission, clutch or linkage you have, the key factor to preventing failure is to make sure all adjustments are correct and maintenance is performed regularly. On a regular basis, check all clearances and check for wear on all moving parts. Keep fluids changes regularly, and inspect for metal shavings or debris in the fluids or grease. This can be a sign of impending parts failure. Check any rod end bearings in linkages for wear, and keep them lubricated.

Driveshaft Selection

The driveshaft is a very important part of the drive line because it delivers the engine power that is turned through the transmission to the rear end. If this particular part is not correct, you will have consistent problems.

Steel Driveshafts

The steel driveshaft has been around longer than any other. In many classes of racing, it is a rule that you have to use a steel shaft. This is a good driveshaft. They are heavier than other types, but they are durable. I remember when we were racing in the early 1980s and lightweight parts were becoming a lot more common. There were not a lot of aluminum driveshafts yet but they were trying to lighten up the steel ones. We were seeing some 3.5 to 4-inch diameter, thin wall driveshafts, but pretty soon they had thinned down to 2.5-inch or 2-inch diameter. As you can see, they were lightening them up by using a larger diameter with a thinner tubing wall thickness. This approach created driveshafts that held up fine. But when they went to a smaller diameter driveshaft, they had to have a thicker wall tubing to keep them from twisting. So they met a happy medium and enlarged the diameter to the 3 to 3.5-inch range and found a good lightweight reliable steel shaft. The smaller 2-inch diameter shafts required a .120-inch wall thickness versus the 3-inch shaft which could use a .083-inch wall thickness. This particular shaft seems to be the most common shaft used today, and is standard on all of our IMCA cars. Driveshaft tubing needs to be 1020 D.O.M. tubing or 4130 for durability.

Aluminum Driveshafts

Aluminum driveshafts have basically taken over late model dirt racing. I would say that ninety percent of the cars that race on the circuit use aluminum driveshafts. They are very light and durable. For the money versus the weight, I don't think you can find a better package. Most aluminum shafts are a 3-inch diameter and have a wall thickness of .250-inch. On an average, they will save about eight pounds of rotating weight over steel. There are several nice aluminum shafts on the market. We use the Coleman shaft regularly. If the rules in your class let you run aluminum driveshafts, then for the money you can't beat the weight savings and durability here.

Carbon Fiber Driveshafts

Carbon fiber shafts are the most expensive type of driveshaft that I have seen, but they are very light. The largest advantage is the reduction of harmonics (which is high RPM vibration and shaft whip) and much lower rotating weight. The carbon fiber material actually absorbs vibrations, so it is easier on other drive line components. ACPT is the company that builds carbon fiber driveshafts. We have run and tested the carbon shaft with Freddy Smith and Bill Frye, and have found them to perform well. The light rotating weight helped gain an additional 200 RPM, so the car could use a slightly taller gear at less RPM.

Because of the nature of the material of these shafts, they won't tolerate a minor impact or nick like an aluminum or steel shaft. The carbon fiber shafts have to be protected from any rubbing or interference. As long as they don't sustain any damage, they are the lightest rotating weight shaft money can buy.

One safety feature about these shafts is that, if you have a driveshaft failure and it comes out of the car, the carbon shaft disintegrates. It won't slap around hitting the driveshaft tunnel or pole vault the car when the shaft digs into the ground. They don't tear as much equipment up and are not nearly as dangerous as a metal shaft. In case of breakage, this makes its cost justifiable. This driveshaft is not going to be the difference between winning and losing, but for the racer that wants every conceivable advantage, this is a good driveshaft.

Determining Driveshaft Length

We always check our driveshafts for length when the car is setting at actual ride height and the car is just like it is going to be raced. Push the transmission slide yoke all the way into the transmission. Then pull it out 1.5 to 2 inches. Very carefully measure from the center of the slip yoke to the center of the rear end flange yoke. This will be the required length of the driveshaft. Our particular driveshaft length with a Bert transmission is 39.5 inches measuring center to center on the u-joints. A Brinn or standard type transmission driveshaft length would be 37 inches.

Once you have installed the driveshaft, check it to make sure there are no clearance problems. Remove the rear shocks and springs and move the rear end housing through all of its suspension movements. Go to extremes to make sure the driveshaft does not bottom or top out. If you have some type of torque arm assembly, then allow for the rear end wrap-up also. Normally, you would look for three inches of torque arm travel at the front of it, but also allow for an additional inch of travel for safety. Check for u-joint clearance through a full range of travel.

Balancing The Driveshaft

Balancing a driveshaft is a required procedure to prevent unwanted vibration. This should be done by a good drive line shop, or make sure that the driveshaft manufacturer has properly balanced it. You don't need vibrations here because this will lead to other drive line related failures.

Checking for trueness is also important. You can do this yourself by putting the car on stands, then put the car in neutral and rotate the rear wheels. While you are turning the wheels, observe the shaft and see if it appears to be out-of-round or wobbling. If you see any problem here the shaft should be replaced or taken to a drive line shop for repair. Ideally, the driveshaft tube should have no more than .004-inch runout.

Universal Joints

Universal joints take a terrible beating in racing, especially on short tracks where they encounter acceleration forces at least twice a lap. To insure the reliability of these parts, it only makes sense to use the best proven product for the job. Dana/Spicer 1310 series universal joints seem to be one of the best choices for a racing application. They are very reliable and affordable.

One of the key factors for u-joint longevity is to make sure they stay well lubricated. Another important factor is to make sure the seals don't have any damage which would let dirt or grit work into the roller bearings. Regularly check side play in the

IMCA and UMP modified rules call for a .25-inch thick by 2-inch wide steel driveshaft loop.

u-joints, and replace them if they show signs of wear. Always use the heavy duty universal joints.

Driveshaft Loops

Driveshaft safety loops are mandatory in almost all stock car racing associations. They are a very important safety item, and they have saved many race car drivers' legs from serious injuries over the years. The purpose of the driveshaft loops is to contain the driveshaft in case of a shaft or u-joint failure.

Most racing associations have a required driveshaft loop specification. IMCA and UMP modified rules call for a .25-inch thick by 2-inch wide steel material loop placed no further back than 6 inches from the front of the driveshaft. Both associations also require that the driveshaft be painted white. This is to make it more visible to drivers should the shaft come out of the car and land on the race track. Other associations require a driveshaft loop of different sizes, and some require two loops, placed six inches behind the front and 6 inches forward of the rear of the driveshaft. This is good safety protection.

At GRT, we go one step further past the requirements and build our driveshaft tunnels out of heavy gauge steel. This provides even greater protection for the driver.

The Rear End

There are two different types of rear ends to choose from – a stock passenger car type or a quick change. For all-out racing performance, the quick change is the only type of rear end to consider. But in many racing classes, such as street stock, hobby stock, limited sportsman, IMCA modified, etc., the rules limit the choice to a stock type of rear end.

The Ford 9-Inch Rear End

The Ford 9-inch rear end is the best choice for oval track racing when track rules limit the choice to a stock type. This is because it is still widely available in wrecking yards, it is lighter in weight than other popular choices, it has very high strength, and it is easy to maintain. The Ford 9-inch, even in its smallest form, has larger axles than the 10-bolt Chevy, and the Ford 9-inch gears and carriers are stronger than the Chevy 12-bolt rearend. And, the gears and bearings are beefier and stronger than the Chevy 12-bolt and 10-bolt rearends.

There have been many hundreds of different applications of the Ford 9-inch rear end through the years. The Ford 9-inch was produced, in all its various forms, from 1957 through 1986. They have appeared with 28 and 31 spline axles. The third member cases have been made from different materials and in different configurations.

The Ford 9-inch has a housing with a removable third member section which allows you to set up your gears on the bench and not in the rear end housing. And, you can carry as many rear end gear ratios as needed because the removable third member makes gear changing relatively easy.

These rear ends have many parts available for them such as a variety of gear ratios and several different spool combinations (like a mini spool). We build Ford 9-inch housings in-house and can put any type of snout on the housing for whatever type of hub you will be running.

Quick Change Rear Ends

Quick change rear ends are used in all types of short track racing. They have come a long way with refinements on these types of rear ends. When I started racing in 1982, just having a quick change

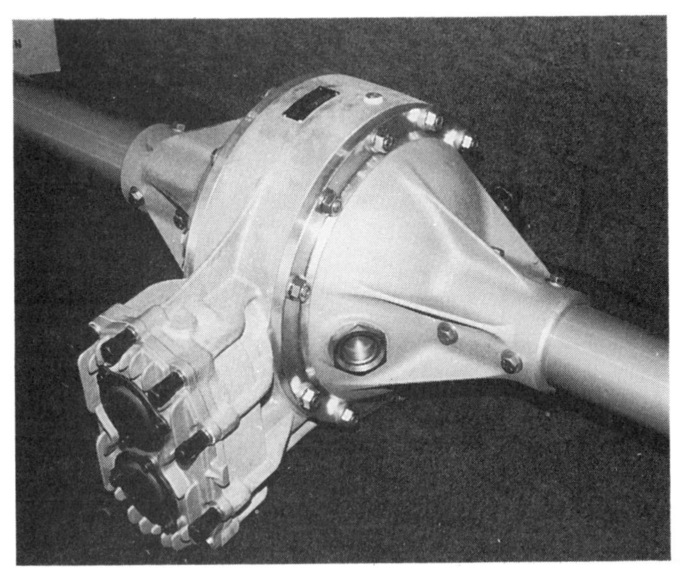
This quick change from Winters features a 6-ribbed side bell housing and a sprint car style gear cover.

rear end was the ticket. Now, to be competitive you have to have all the bells and whistles.

The first quick changes were just aluminum eight-sided side bells and housings with a large heavy gear cover. They had steel tubes and were bulky and heavy. Now, quick changes which are used under dirt late models are extremely light. They have aluminum tubes, magnesium side bells, and center sections with sprint car gear covers. The sprint car style of gear cover is a flat cover which seals off the quick change gears that are housed in the center section. A regular quick change gear cover wraps all the way around the gears and has long bolts that attach it. This is a much heavier piece than the sprint car cover.

The current style of quick changes have hardened and drilled lower shafts, lightweight aluminum spools, inspection holes in the bells to check ring and pinion wear, and even titanium bolt kit options are available. These rears are extremely light (about 70 pounds) and very dependable. Every car we build that leaves with a rear end has the aluminum tubes, magnesium side bells and center section with a sprint car gear cover. We also use the lightweight aluminum spool as standard equipment. You can't get a lighter, more dependable rear end than this. There are several manufacturers of quick change rear ends and they all are good, but we use Winters rear ends 95% of the time.

We have tried to go even lighter with what they call a mini quick change. These are manufactured for a mini stock application. But, they just won't hold up in a full sized stock car. The mini quick change is good for what it is designed to do, but it is not recommended for a dirt late model.

It is advisable to have a complete set of spur gears for the quick change. This allows you to have every possible gear ratio available. You can change these at the track in just a few minutes by taking off the rear cover. You usually loose about one quart of grease when making a gear change. We dispose of the grease that we drain in making gear changes and add a fresh quart. This helps to keep fresh grease in the rear end. And, it gives you an opportunity to inspect the old grease during changes to determine if any rear end failure is starting to take place. When putting the grease back in the rear end after a change, we use what they call a fill can that holds about one quart of grease. The fill can has a hose running from the bottom of it to the rear end center section so the grease will be draining into the rear end as soon as you pour it in the can. What this means is that once the crewman who is changing the gears has installed the cover, even while he is tightening the bolts another crew member can be pouring in the replacement grease. This speeds up the process. Once the bolts are tight on the rear end cover, you are back racing again without having to wait on rear end grease installation.

Axles

There are two different types of axles that can be used – a solid axle, and a gundrilled axle. The gundrilled axle has a hole drilled all the way through the center core that lightens the axle considerably. We use the gundrilled axle exclusively. For the weight loss versus the cost over a solid axle, it is by far the best deal. On top of this they are very reliable. For a cost of only $40 per side you will save four pounds per side. That is eight pounds of rotating unsprung weight being saved. That saving is worth every penny.

You can take it one step further and install a titanium axle, which saves another two pounds per side. But, here is where the expense come in. Titanium axles will be about $300 more per side, so this is where you will have to decide if you want the weight savings versus the extra cost.

Drive Flanges

We finish off the lightweight rear end assembly with a magnesium hub and a light weight aluminum drive flange. There are several different types of drive flanges, and we have tried all of them. Winters makes a cushioned drive flange that has rubber inserts that come in different hardnesses. These inserts are used to cushion drive line torque.

At first we thought this would be an advantage to cushion tire spin but found that we gained nothing in performance or times. But, we gained by reducing drive line related failures. This is because the rubber inserts in the drive flange cushion and absorb the initial shock for the axles, yokes, u-joints, driveshaft and all related drive line parts. The only drawback is the added maintenance on the drive flange. The rubber inserts fatigue quickly and they require lubrication regularly. They require extra maintenance, but they reduce drive line component wear. We use regular drive flanges as standard equipment, but we offer these drive flanges as an option.

Differential Types

There are several different types of differentials available today for rear ends. We have probably used every kind there is at one time or another, but in our opinion, late model dirt cars work best with a standard spool. I'm not saying that there is not a place for other differentials, but let me explain why we think spools are best.

With a spool you always know what the rear end is doing. It is locked up all the time and any of the chassis adjustments you make won't be affected by the rear end. Let's say the differential is not working properly and you don't know it. You make all kinds of adjustments to the chassis and the whole time it is the rear end is not working properly. All of your adjustments have been wasted. With a spool, you know that if you make a change to the race car, it works or it doesn't work, and a differential had nothing to do with the adjustment. With the various types of lockers or torque bias differentials, you always have a variable that could be the problem, but you never know for sure. On the other hand, a

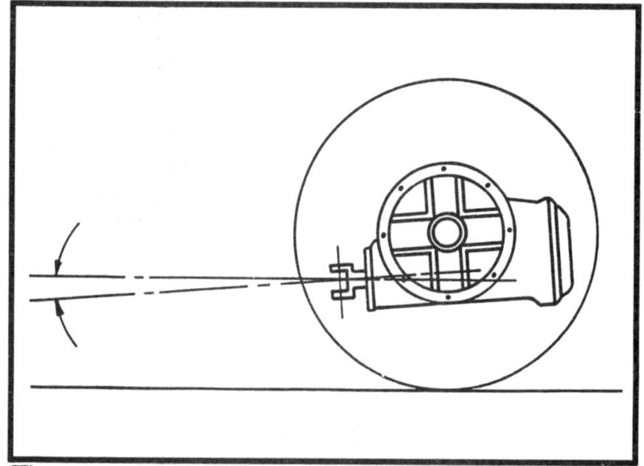

The rear end is set with a negative pinion angle to prevent the pinion from over-centering during travel.

differential or locker offers advantages with varying types of track conditions.

With a Detroit Locker-equipped rear end, the inside rear wheel is disengaged and it rotates freely at corner entry. Under power, the Detroit Locker solidly locks both wheels together. The advantages of using it are eliminating corner entry understeer, and reducing the amount of tire stagger required to turn the car.

With a locker, you have mechanical parts that can wear or fail and this is the reason I have never felt comfortable with this type of differential. Lockers also add unsprung weight to the car.

A torque bias differential does not have as many mechanical parts as a locker. A locker has springs and cog type gears that act like a ratchet, whereas a torque bias differential just has gears that have close tolerances which produce the free wheel effect on corner entry. From what I have seen, the gears in those units can wear and this will prevent it from functioning properly. Keeping them properly maintained is a key factor.

Provided the differentials are working properly, the theory is that you can free wheel or let the car turn through the corner easier whereas a spool locks the axles together solidly and relies on stagger alone to turn the car. So as you can see the torque bias differential would let you use a little less stagger. Provided the torque bias differential works properly, you can free up your car into and through the apex of the corner. Just make sure that if you run that type of differential, it is well maintained.

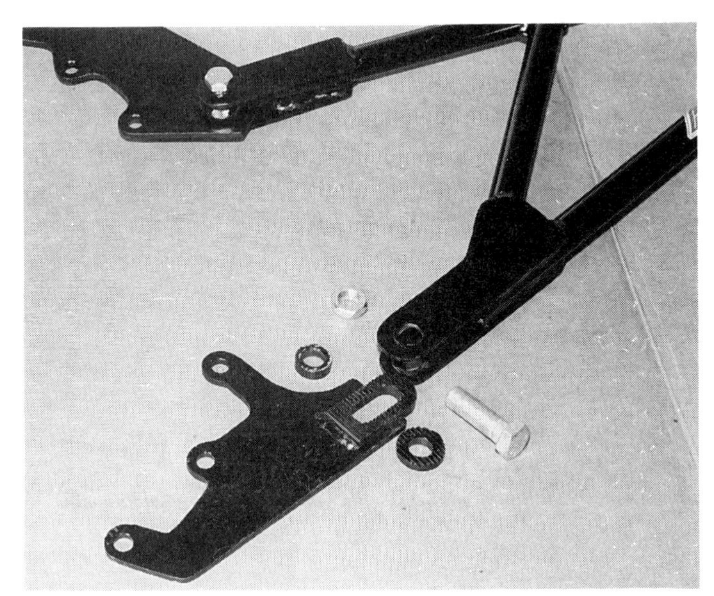

This new adjustable torque arm from AFCO allows adjustment of the arm to set 3 to 8 degrees of pinion angle and still keep the arm parallel to the frame rails.

Setting The Pinion Angle

Pinion angle is a topic often discussed that has different opinions. I've seen several different angles on different kinds of cars. The best way to tell what you need is to check the pinion wrap-up under torque. The front of the torque arm on our race cars will travel upward 3 inches at 38 inches from the rear axle centerline. This is the maximum amount of travel that we recommend. This will enable you to see the amount that the pinion travels. We don't like to see the pinion actually over-center or start going up. Our normal pinion angle setting is six degrees down and this amount prevents over-centering from happening. From what we have seen, anywhere from 5 degrees to 7 degrees down is a good place to start. If you have a solid link type of rear suspension (such as a 3-point) that has no wrap up movement, less pinion angle would be appropriate — generally in the range of 2 degrees to 3.5 degrees down.

The pinion angle is set to operate in a narrow angularity range in order to create the most efficiency from the universal joints. U-joints are intended to operate at a slight angle in order to preload the roller bearings. But more than normal misalignment causes high stress and wear on the u-joints.

As far as actually adjusting the pinion angle, it depends on what type of torque arm or fifth coil set up you have. Our particular cars use a fifth coil torque arm with a rebound chain that has a specific length. When that is combined with the torque arm design, it automatically sets the pinion at 6 degrees down. The torque arm chain determines the static pinion angle. Cars with a third link or top link suspension usually set pinion angle with the length of the bar and usually use rubber biscuits that act as a rebound cushion. So the rubber biscuits can be used as the adjustment to determine the actual static pinion angle. Pinion angle is important so make sure you have this on your check list.

Gear Lube

There are many gear lubes available. We use a special lubricant called Royal Purple. This is a synthetic grease. It has improved wear and increased reliability for us. It also does not have to be changed as often as a regular mineral-based grease and therefore this offsets the cost difference. Our rear ends seem to run cooler and are quieter with this lubricant. Regular gear lube like Valvoline or Pennzoil, etc. in the 80-85 weight range, is also a good basic grease. This lubricant needs to be changed more frequently than the synthetic, and it doesn't seem to run as cool as the Royal Purple. In general the synthetic lubricants have performed flawlessly and we recommend using them.

Proper Gearing

The car must be geared for the size of the track so that the engine maximizes its power range. If your car's power range is peaked at 7,500 RPM, gear the car so that 7,500 RPM is reached when you need to lift for corner entry.

Don't gear the car so the engine bogs down at corner exit thinking that this will help you keep the tires from spinning. This hurts performance in two ways. First, the car is lugging and not delivering the power that it could. When you get to a point on the race track that is slick, the tires will act like you are on ice. It is hard to control throttle response.

Secondly, gearing a car too high will hurt braking. A properly geared car will let the gear help slow it on corner entry and therefore you use less braking. We have learned through many races and test sessions that leaning toward the lower gear ratio provides better performance all the way around. Gear the car for the size of the track and the engine's power range versus its RPM range.

Chapter
14

Chassis Set Up & Alignment

Before we set up the chassis, we have to make sure that there is no binding in the suspension, and that the rear end is set square in the car. If any of these problems are present in the chassis, they will negate any of the settings we try to make on the car.

Checking For Chassis Binds

Before doing the chassis set-up in the shop, you must look for the presence of any chassis binds in your completed car. Move each of the wheels through at least two inches more than their normal wheel travel. Carefully observe the movement of everything attached to that wheel. Look for shock absorber binding or bottoming out, A-arms moving freely or contacting the frame, the steering shaft moving freely without contacting anything when turned , a free movement of all steering components through full range of left-to-right steering with no binding or contacting, the Panhard bar moving freely with no binds or without contacting any chassis parts, the torque arm moving smoothly with no contact or bind, and the rear suspension arms moving freely with no binds.

If you observe any problems, be sure to correct them right away before proceeding to the chassis set-up.

Squaring The Rear End

Simply, the rear end squaring process is making sure the rear end housing is set straight in the car — perpendicular to the vehicle centerline and not angled. If the right rear is set behind the left rear, the car will be loose. If the right rear is set ahead of the left rear, it will push.

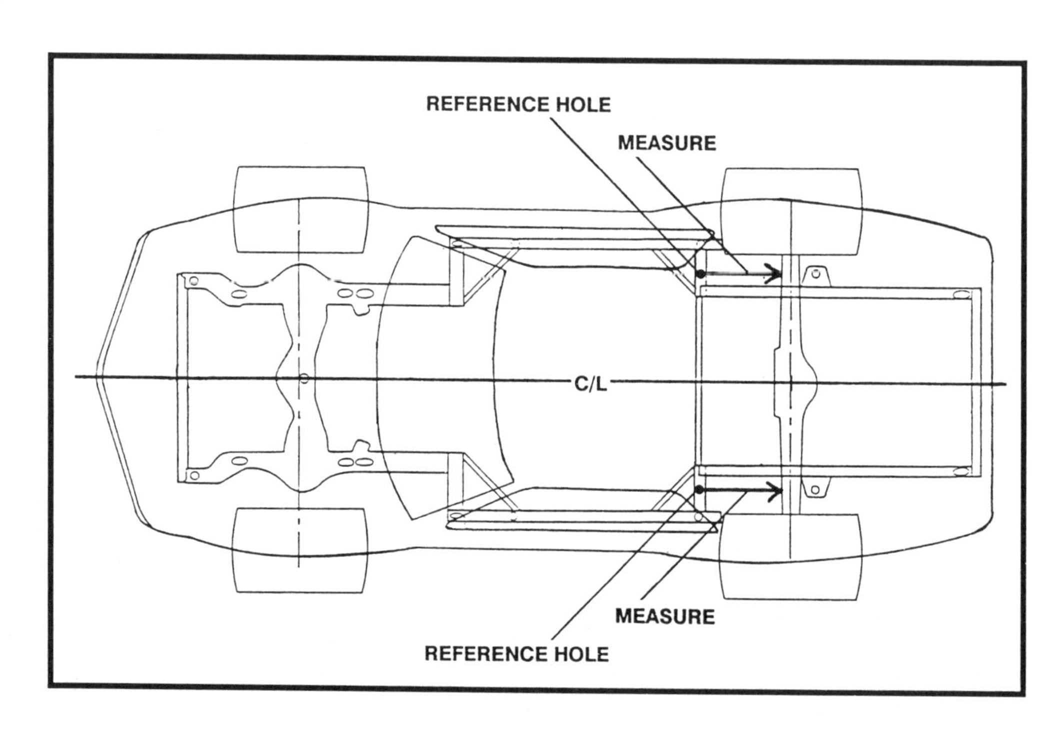

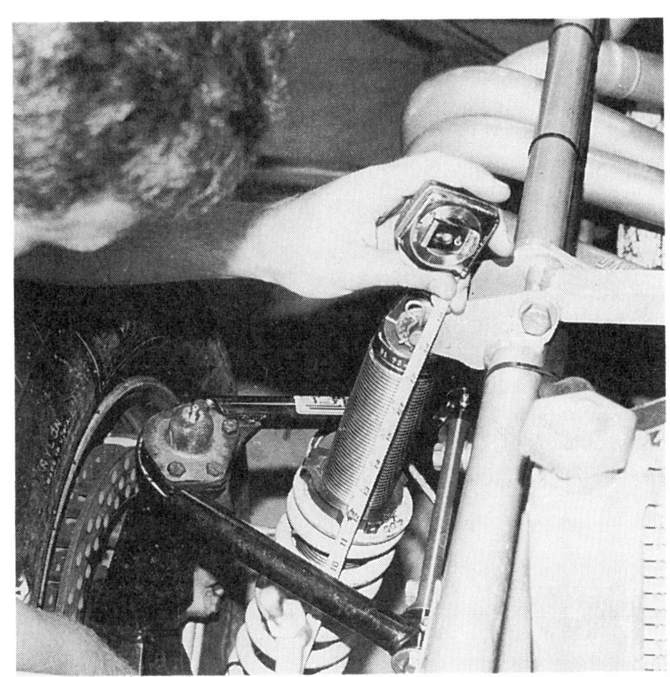

A quick and positive way to set the ride height is to measure from the top of the lower A-arm to the bottom of the main frame.

Squaring the rear end is very critical to the car's handling. Even a 0.25-inch out-of-square can have a significant effect on handling.

Most professionally manufactured chassis have built-in squaring reference marks on the frame rails. These reference marks are usually holes drilled in the frame rail, or marks punched in the rails. Make sure you know where these are. If you have built your own chassis, make sure you remember to include these squaring references on the frame rails.

On our cars, we square everything off of the right side main lower frame rail. This frame rail is the only one in the car that is 90 degrees to the rear axle. We square the rear end 90 degrees to this frame rail, having the rear end at proper ride height. Then we measure to a point on the frame straight in front of the rear end housing and center punch a mark. This measurement will always serve as a reference point for checking squareness of the rear.

Also, our four-link brackets are square and if you put the radius rods at the specified length, then you know the rear end is square.

Chassis Set-Up In The Shop

Your goal in setting up the chassis at the shop is to have the car ready to race competitively as soon as it rolls off the trailer at the track. With the car set up properly at the shop, you should have to make a very minimum amount of adjustments at the track.

Be sure that you choose a flat, level surface in your shop on which to do the set-up. Always use the same place. Make marks on the floor where the car sets so it can be returned to the same location time after time.

The chassis set-up should follow a specific order each time. Make sure the correct springs and shocks are in the car, and that it is completely race-ready. All fluids (fuel, oil, water, power steering fluid, transmission fluid, etc.) should be full, and the wheels and tires (including tire stagger) and air pressure should be the same as you plan to use at the track. Have the car prepared just like you are going to race it. We have the car fueled with 15 gallons of fuel because this is the median point between the start of the race and the end of the race.

Setting The Ride Height

The first step in chassis set up is to establish the recommended ride heights and front end alignment. Start with setting the ride heights at the manufacturer's recommended heights. We provide ride heights and a complete set up sheet with each car we sell.

A quick and positive way to set the ride height is to use a set distance between two reference points. We measure from the top of the lower A-arm to the bottom of the main frame. Our cars require this distance to be 3.125 inches on both the left and right front. Some manufacturers might require you to maintain a certain upper A-arm angle, but we find it is easier to use the reference distance measurement. Adjust the spring seats on the coil-overs to obtain the required measurement. Be sure to bounce the car up and down thoroughly to settle the suspension and the coil-overs. Double check the reference measurements one more time after you do this.

Once you have adjusted the two front ride heights, go to the rear. The right rear ride height is measured between the top of the lower underslung frame and the bottom of the rear end tube. 3 inches is our standard measurement for GRT cars at the right rear. An advantage with our cars having underslung frame rails is that it allows you to use ride height blocks between the rear end tubes and the lower frame rails.

Every time you make a suspension setting change, bounce the car up and down thoroughly to settle the suspension and the coil-overs.

These blocks set the proper ride height distance between these two points.

The rear settings are a little different due to the fact that you only need to set the ride height on the right side. Let me explain. On any race car I have ever set up, you only set three corners – the two fronts and the right rear. The reason you leave the left rear out is because you have to set the wedge with that corner. The wedge is how much left rear corner weight you have compared to the right rear corner. If you had all four ride heights set to an exact measurement or angle, there would be no way to change the wheel weights to get what you want. So, the left rear ride height is allowed to be whatever it takes (within reasonable tolerances) in order to get the corner weights of the car set correctly.

Setting The Corner Weights

Once you set the three corners and look at the three wheel scales, you can determine what adjustments have to be made for the left rear. Let's say you have 550 pounds on the left front and 450 pounds on the right front. The right rear is 600 pounds and the left rear is 550 pounds. This gives you 50 pounds of right rear weight or reverse wedge. Your ride heights might be exactly what they are supposed to be, but you want to have 50 pounds of left rear weight or wedge. The goal here is to have the left rear weigh 50 pounds more than the right rear.

Having underslung frame rails allows you to use ride height blocks to set the proper ride height distance.

Using grain scales to scale the car gets the car up 8 inches off the floor for the convenience of working on the chassis.

Now comes the part that a lot of people have a hard time understanding. If you move the weight jacks exactly opposite of each other across the front and across the rear, the ride heights will be maintained but static weight will be moved around. For example, we are starting with 550 pounds at the left front, 450 pounds at the right front, 550 pounds at the left rear, and 600 pounds at the right rear, and the ride heights are correct on each front corner and on the right rear. We want to change the weight around without changing the ride heights.

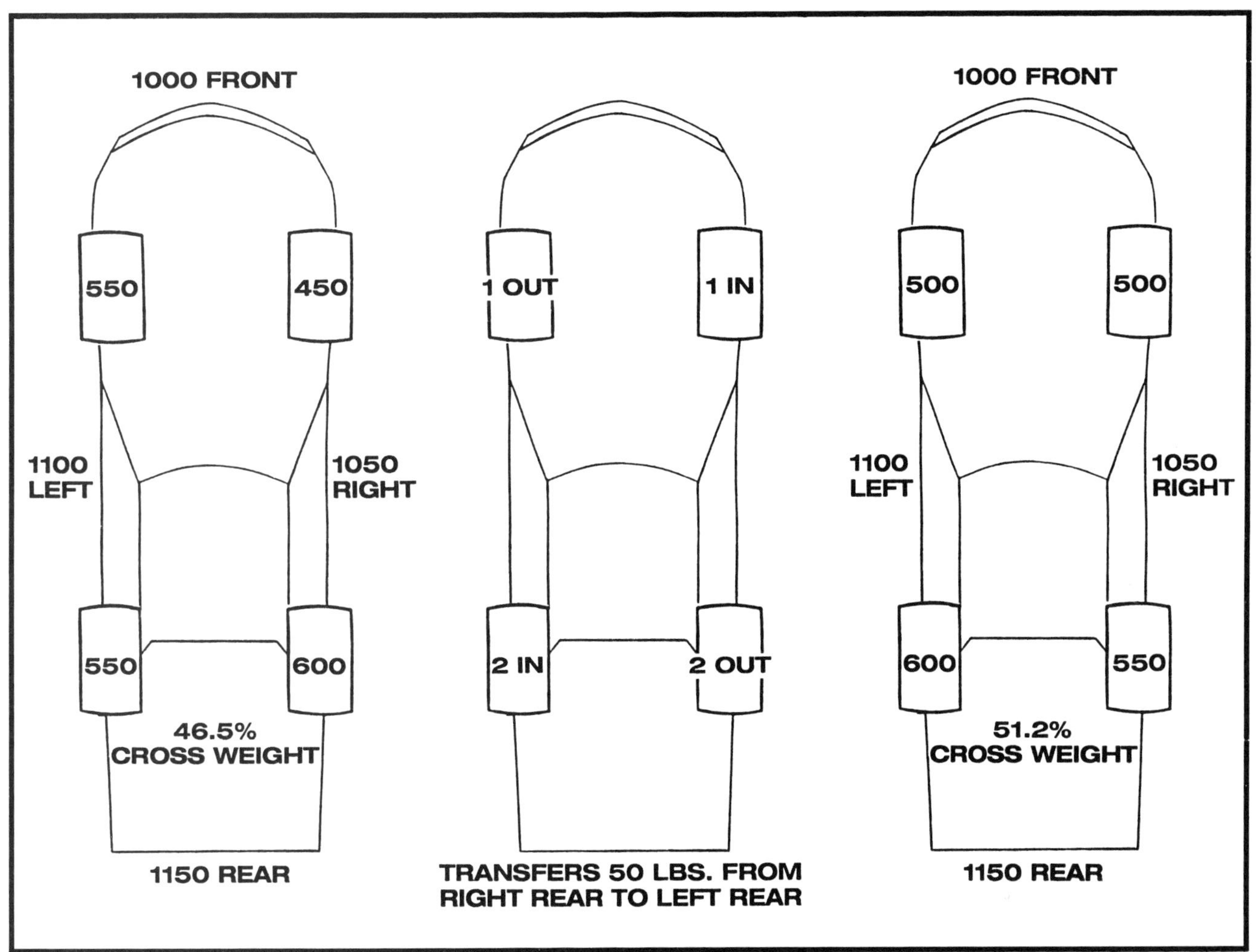

This weight jacking method transfers 50 pounds of weight from the right rear to the left rear without changing chassis ride heights. Notice what happens to the front corner weights after the change. Also notice that the total of the left side, right side, front and rear weights did not change even though the corner weights did.

First let me explain the system I use when scaling a car. Any time you want to change 50 pounds of weight around, it is a simple system to use. Put two turns down on the left rear coil-over and turn the right rear coil-over adjuster up two turns. At the front, turn the right front coil-over adjuster down one turn. At the left front, take one turn off of the coil-over adjuster (move it up). After these adjustments to the coil-overs are made, be sure to bounce the car up and down thoroughly to settle the suspension and the coil-overs, then re-weigh the car.

With these moves, you have transferred 50 pounds of weight from the right rear to the left rear. Now you will have 550 pounds on the right rear and 600 pounds on the left rear. And most importantly, the existing ride heights will have stayed the same. You will also find that this system has moved 50 pounds from the left front to the right front, so that now both front corner weights are 500 pounds. (Note: This system might vary in its results slightly from one car to the next and may not come out exactly the same, but experiment with it on your chassis. This system works.)

Study these changes so you know what each adjustment does. To take weight off of a wheel, turn the weight jack or coil-over adjuster away from the spring. When you do this it also takes weight away from the diagonally opposite corner of the car. If you take weight off of the left rear, you also take weight off of the right front.

If you turn the weight jack adjuster toward the spring, you will add weight to that wheel and at the same time add weight to the diagonally opposite

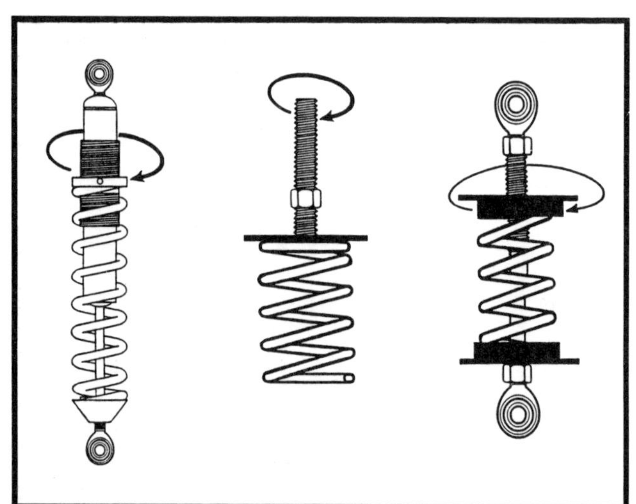

Whether you are working with a coi-over unit, a weight jacker over a coil spring, or a coil spring slider, the weight jacking adjustment method is all the same. If you turn the weight jack adjuster toward the spring, you will add weight to that wheel and at the same time add weight to the diagonally opposite corner.

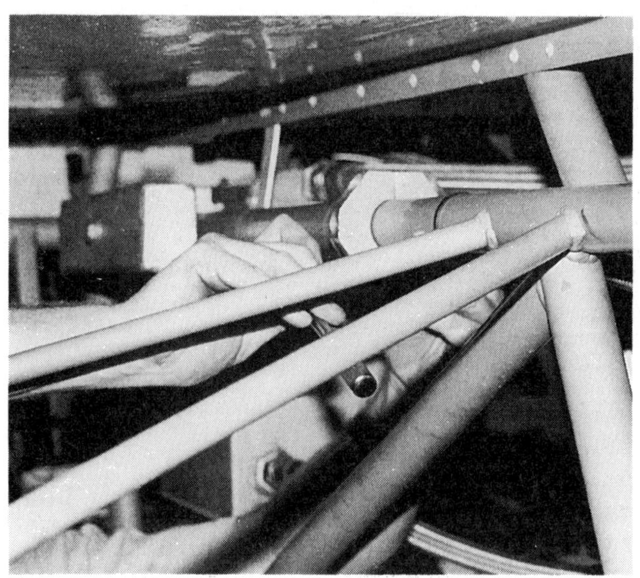

The higher that weight is placed in the chassis, the more overturning moment there will be during cornering. This promotes side bite or right side traction.

corner. For instance, adding weight at the left rear adds weight at the right front also.

Making weight adjustments on the front springs of a race car will have a greater effect than making the same adjustment on the rear springs. This is because the front springs are stiffer rate springs. One turn of a weight jacker on a 500#/" spring will jack more weight than one turn of a weight jacker on a 225#/" spring. This occurs because a stiffer spring offers more resistance to the movement of the weight jacker, thus the chassis raises or lowers more quickly than it would over a softer spring.

Remember, to start this system you must first get the ride heights correct, and then scale the car to see where it is at. Once you know where you are, you can start moving static weight around.

Another consideration is that the left side weight percentage and the rear weight percentage will not change, no matter how you adjust the static weight by raising or lowering corners of the car. The left, right, front and rear corner weights will always add up to the same number regardless of what changes are made to each corner weight by adjusting corner heights. The only way to change one of these percentages is to add or subtract weight to the car or change wheel offsets. So, for instance, if you scale your car and you have 53% left side weight and you want 54% left, you must add ballast or move components around, or change wheel offsets. We do not recommend changing the wheel offset on our cars to change a percentage, so you must add ballast to the most practical area. Make sure that when you add ballast, the weight is securely bolted to the chassis by a ballast clamp or to a welded insert in the frame.

Ballast Placement

The height of the ballast in the chassis has an effect on the handling of a car more so than just what the scales say. For instance, having 54 percent left side weight that uses 75 pounds of weight mounted low in the chassis will react differently than 75 pounds that is mounted high. So always take this into consideration when mounting ballast.

Ballast location should be considered for all track conditions. With dryer smoother tracks, place the ballast higher in the chassis. For rough, hooked up tracks, the ballast should be placed lower in the chassis. The higher that weight is placed in the chassis, the more overturning moment there will be during cornering. The overturning moment promotes side bite or right side traction during cornering. Heavy or hooked-up tracks create a lot of side bite for the tires, so the car does not need additional overturning moment. However, dry slick tracks create very little side bite, so more overturning moment is required for right side traction. So, ballast is placed

The track surface condition influences how the weight percentages are set for the chassis. A track that stays "hooked up" (good moisture and traction) will require more left side weight percentage. Photo by Dennis Mattish

higher in the chassis on these tracks to create the desired affect.

Once ballast is added to the chassis, the car most likely will settle and ride heights will have to be adjusted accordingly. We also recommend that the caster and camber be set before final scale readings are made.

Setting Up For The End Of The Race

The most common way to set a car up for the final scale reading is to have the chassis set up with the fuel load the way that it will be at the finish of a race. The reason is that the minimum weight you can have is when the race is over. So, you want to have the chassis scaled based on the optimum percentages with the car at its minimum weight.

When a car is set up to be neutral handling at the end of the race with a nearly empty fuel load, that means the car will start a race with an understeering condition when the fuel cell is full. The driver may need to use brake bias adjustments to help the handling until the car starts to neutralize. The car will get looser as the fuel burns off, so a compromise has to be made for the set-up at the beginning of the race. You don't want the car to be loose at the end of the race.

The track surface condition will influence how the weight percentages are set for the chassis. A track that is anticipated to stay "hooked up" (good moisture and traction) will require more left side weight percentage. However, if the track will be dry slick at the end of the race, more rear weight percentage (more rear ballast) would be required.

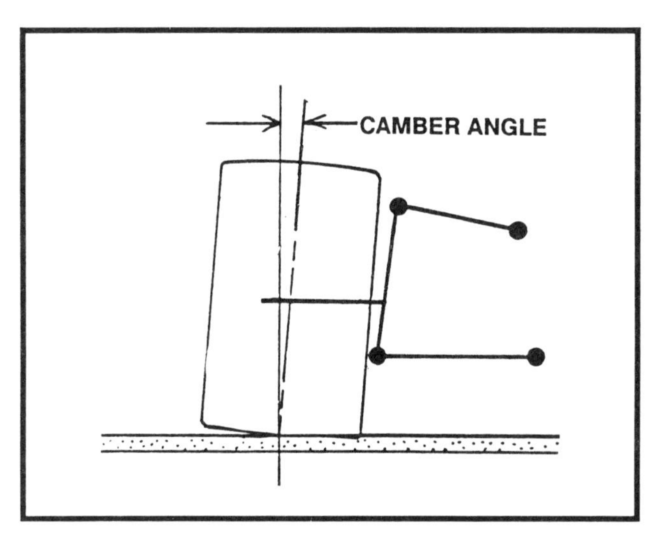

Don't get the chassis too light. Remember to use enough ballast in the chassis to get it to minimum weight with a very low fuel load. Give yourself some room here so you don't get disqualified for being too light after a race.

Front End Alignment

The front end alignment is the process of setting the desired camber and caster at each front wheel, plus setting the toe-out.

Camber is the inward or outward tilt of the wheel in the vertical plane. Negative camber tilts the top of the wheel inward toward the centerline of the car. Positive camber tilts it outward away from the centerline of the car.

Caster is viewed from the side of the wheel. Positive caster places the top ball joint behind the center of the lower ball joint. Negative caster is the opposite. Negative caster is rarely used in stock car chassis settings.

Camber, caster, and toe-out are very important settings for the proper handling of a race car and should be checked often. These affect tire wear and steering stability. Incorrect toe-out will cause a darting condition. Check these settings before each race. Most probably if the camber, caster, or toe-out is not correct after you have properly set them, you have

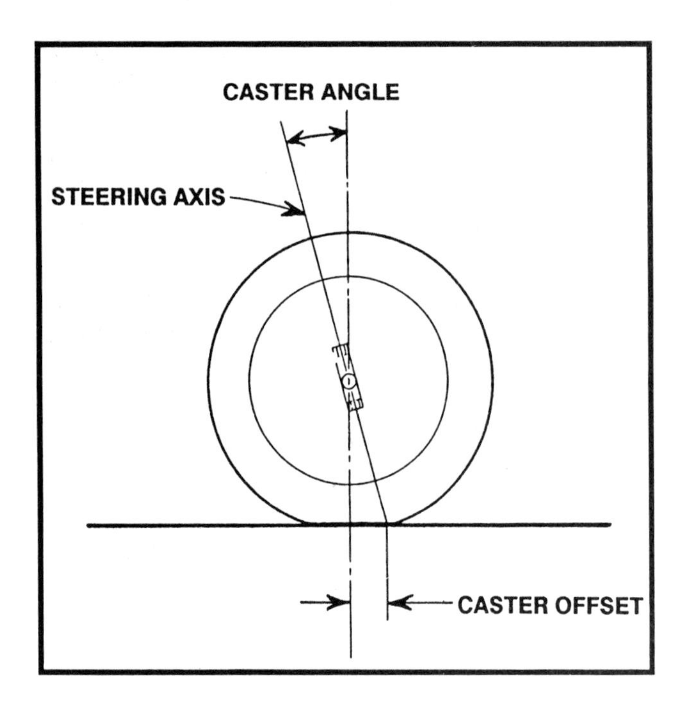

a bent ball joint or some other suspension component which is bent or broken.

Caster

Setting the caster should be done before the camber because camber is affected by caster.

Caster provides directional steering stability. This influence is created with a line which is projected from the steering pivot axis down to the ground. This line strikes the ground in front of the tire contact patch when the caster is set in the positive position. A torque arm then exists between the projected steering axis pivot line and the center of the tire contact patch. The torque arm serves to force the wheel in a straight ahead direction. The greater the length of this torque arm (caused by greater amounts of positive caster), the greater the steering effort required to turn the wheels away from their straight ahead direction.

The difference in caster setting between the left front and right front is called caster stagger. A slight amount of caster stagger helps the car change from its straight ahead path more easily to ease into a left turn. This caster stagger is created by having the caster at the right front greater than the caster at the left front.

The caster stagger on a dirt track car is generally not very large because the same factor which aids in turning the car to the left increases steering effort when turning to the right to countersteer. Most drivers are generally comfortable with a 2-degree caster split between the left front and right front on dirt.

When a wheel with positive caster is steered, it raises the spindle height, and thus jacks weight to the diagonal rear corner.

Positive caster jacks wedge into a chassis. When a particular wheel with positive caster is steered, it raises the spindle height, and thus jacks weight to the diagonal rear corner. The greater the positive caster, the more weight that is transferred. So on a dirt track, more positive caster at the left front means that when the car is countersteered to the right to catch the rear end, more weight is transferred to the right rear to give it a bite.

The amount of caster and caster stagger used on a race car is also influenced by the use of power steering. Cars with power steering can use more positive caster and more caster stagger.

Setting The Caster

On most dirt late models, caster is set by adjusting the lower strut rod back and forth. If you pull the lower A-arm forward with the strut rod, you add caster to the wheel. If you push the wheel back by lengthening the strut rod, you take caster out. Do not adjust the caster with upper A-arms that have a solid shaft in them by using spacers that are not equal. This procedure will cause the cross shaft in the upper A-arm to be in a bind.

Caster is set by adjusting the lower strut rods. Do not adjust the caster with upper A-arms that have a solid shaft in them.

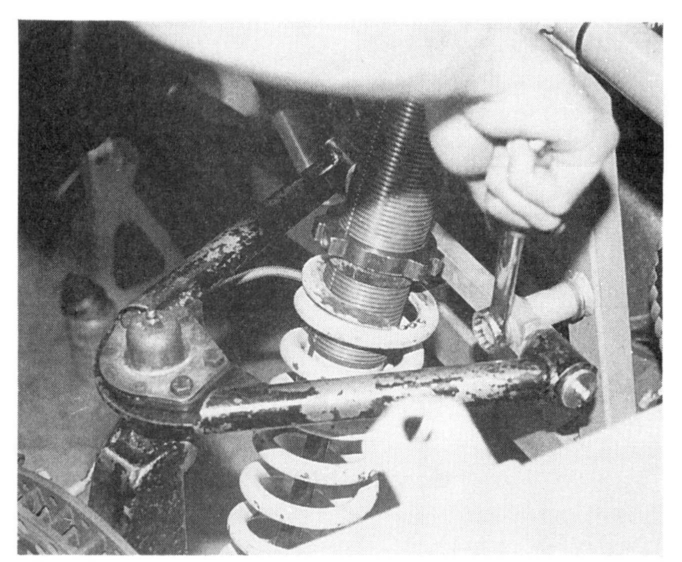

Spacing the upper A-arms in or out adjusts camber in or out.

Camber

The purpose of the camber adjustment is to keep the tire contact patch flat on the track surface at the maximum point of cornering. Camber has the biggest influence over the vehicle's cornering ability than any other alignment feature. On most dirt track late model cars, the static camber setting at the right front tire is between 1.5 and 3 degrees negative, depending on the camber change curve of the suspension, the type of track and banking, the tire construction, and the tire tread width. In general, wider tires use less initial camber, and narrower tires use more.

Spacing the upper A-arms in or out adjusts camber in or out. Moving the upper A-arm in toward the vehicle centerline adds negative camber. Moving it out adds positive camber.

Use the following guidelines for the initial camber settings (track testing may show that initial settings may need to be changed):

Dirt Late Model Car

Left front: +1º to +2º Right front: –1.5º to –3º

Sportsman Car (Over 2,800 lbs.)

Left front: +1.5º to +2.5º Right front: –3º to –3.5º

Note: the difference in the camber settings shown above are caused by the difference in tire widths used by the two classes. The dirt late model uses a 13 to 14-inch wide tire, while a sportsman class car generally uses a 10 to 12-inch wide tire. Other factors listed above also influence the camber selection.

Tire temperatures taken after practice laps will help you determine the exact camber requirements for your application.

Setting The Toe-Out

The range of toe-out used on dirt track cars varies from 1/8-inch to 1/4-inch. More toe-out causes the front end to stick more as the car enters a turn. If the car is sticking too much in the front, use less toe-out. Don't set the front toe-out at more than 1/4-inch. It would cause the front end to scrub off too much speed down the straightaways. Start with an initial setting of 1/8-inch.

Toe-out is set by measuring the difference between the front of the tires and the rear of the tires at spindle centerline. There are several ways to accomplish this, but one of the easiest is to use a trammel bar set solidly against the right front tire. The other end of the bar has a pointer set a few inches away from the left front sidewall. Measure from the pointer on the bar to the sidewall at the same height front and rear on the tire.

When measuring to the sidewalls to find and set toe-out, you have to be very careful about sidewall

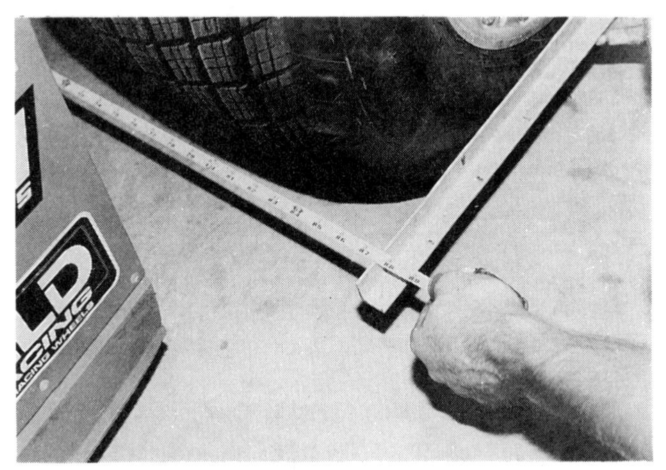

When measuring to the sidewalls to find and set toe-out, you have to be very careful about sidewall distortions.

distortions. Tires can have high and low variations that can easily make your measurements off by at least 1/4-inch. So before measuring for the toe-out, run through this procedure first to assure you have valid reference points:

Jack the wheel up and rotate the tire against a fixed reference point, such as a jack stand. In doing this you are looking for the highest and lowest spot on the sidewall. Mark the highest and lowest spots with chalk, and then put those marks at the top and bottom when you are doing your measurements for toe-out. Every time you make an adjustment on the toe, you have to roll the car back and forth before you re-measure. With the marks at the top and bottom, you know you are getting the same reading every time.

Once an adjustment has been made, roll the car back and forth to take up any slack in the linkage. If the car had to be jacked up to make the change, also bounce the car up and down before taking a new toe measurement.

When setting the toe-out for a dirt track car, don't split it equally between the left front and right front wheels. Set the left front straight ahead and set all of the toe-out in the right front.

Checking the amount of toe-out is a real quick way of determining if anything in the front end has been bent due to contacting something on the track. Be sure you know what the toe-out setting was when the car went out on the track. If the car bangs wheels with another car, or hits something, be sure to check the toe-out right away. If it has changed, chances are something is bent.

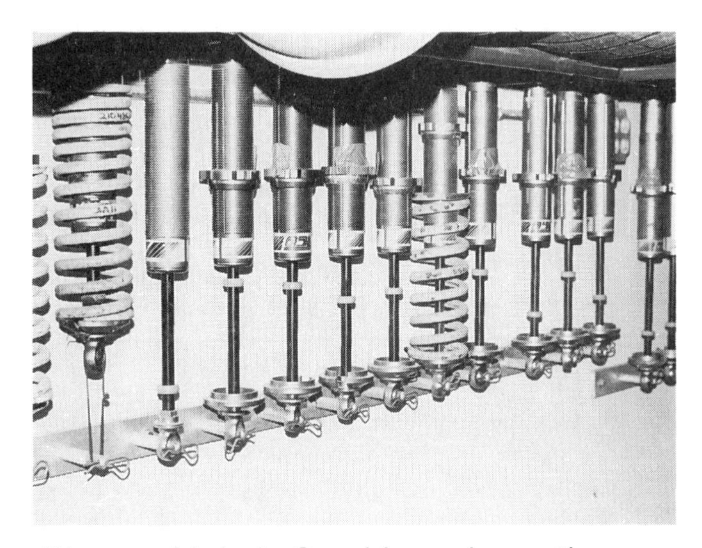

(Above and below) Organizing and presetting your spares saves precious moments at the track. The springs and shocks have already been scaled and set on the car, so they can be bolted on without rescaling. The strut rods and A-arms have also been preset on the car to proper lengths.

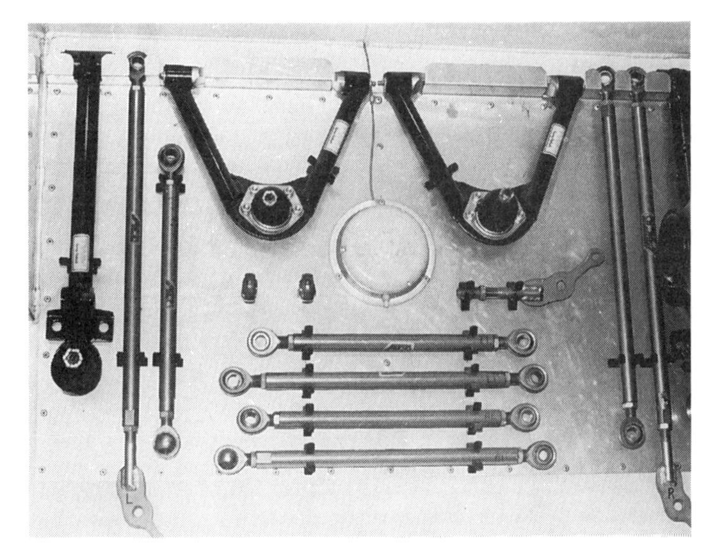

Starting Specs

The **Recommended Starting Specifications** charts show basic guideline starting set-ups for an "average" dirt track and two different types of cars. These numbers are based on a 3/8-mile oval, with up to 10 degrees corner banking. This set-up will work on a 1/4-mile and a 1/2-mile track as well, with probably just a few adjustments.

Recommended Starting Specifications — Late Model Car

Track Type: Flat to med. banked 3/8-mile
Car Weight: Up to 2,500 pounds
Track Condition: Good traction, fairly dry

Weight Distribution

Left: 52% to 53%
Rear: 54%
Cross: 45% to 47%

Spring Rates (Coil-Over)

LF - 450 to 500 RF - 500 to 550
LR - 200 to 225 RR - 225 to 250

Shock Absorbers

LF - 75 RF - 75
LR - 94 RR - 94

Front End Alignment

Caster (power steering)
LF +2° to +3° RF +3° to +5°
Camber
LF + 1° to +2° RF –1.5° to –3°
Toe-out: 0.125-inch

Recommended Starting Specifications — Sportsman Car

Track Type: Flat to med. banked 3/8-mile
Car Weight: 2,800 pounds
Track Condition: Good traction, fairly dry

Weight Distribution

Left: 52% to 53%
Rear: 53% to 54%
Cross: 49% to 50%

Spring Rates (Conventional coils)

LF - 900 to 1,100 RF - 1,400 to 1,800
LR - 200 to 225 RR - 225 to 250

Shock Absorbers

LF - 76 RF - 76 to 77
LR - 94 RR - 95

Front End Alignment

Caster (power steering)
LF +2° to +3° RF +3° to +5°
Camber
LF + 1° to +2° RF –1.5° to –3°
Toe-out: 0.125-inch

Recommended Starting Specifications — IMCA Modified

Track Type: Flat to med. banked 3/8-mile
Car Weight: 2,400 pounds
Track Condition: Good traction, fairly dry

Weight Distribution

Left: 52% to 53%
Rear: 54% to 55%
Cross: 46% to 47%

Spring Rates (With a 3-point rear susp.)

LF - 750 RF - 850
LR - 250 RR - 225

Spring Rates (With a 4-bar rear susp.)

LF - 750 RF - 850
LR - 225 RR - 200

Shock Absorbers

LF - 74 RF - 75
LR - 94 RR - 95

Front End Alignment

Caster (power steering)
LF +2° to +2.5° RF +4° to +4.5°
Camber
LF + .75° RF –2.5°
Toe-out: 0.125-inch

Chassis Setup & Adjustments

The set-up should be done with the car race ready and with 15 gallons of fuel, but with no driver. This is a car with a 32-gallon fuel cell, and which starts a race with a full fuel load. If you start the race with a lesser fuel load, scale the car with 50 percent of the fuel load you start a race with. We usually base most settings on an average of a 180-pound driver. If the driver weighs more or less, then allow for this in your calculations. For example, if the driver weighs 225 pounds, then the left side percentage would be less because of the additional driver weight. The main thing about setting your car up is do it the same way every time and in the same place.

Let's say that you are going to a track that has a cushion or is extremely rough. Then you need to plan on stiffening the right side of the car. Shocks should be increased to a 76 at the right front and a 95 at the right rear. Spring rates will usually increase 50 pounds in the front and 25 pounds in the rear.

You can judge a lot of this by your shock indicator travel, which is a good reference.

Shock absorber travel on the right side of the car will give you a good guideline for proper chassis performance. On our cars we look for 2.75 to 3 inches of shock compression travel on the right front and right rear. The shock travel would be the same for a wet, muddy track, a moist hooked-up track and a dry slick track. The difference would be in the shocks and springs used in the car for these different track conditions to maintain this travel.

Extremely smooth, dry race tracks generally will require a little softer springs and shocks. And, possibly split valve shocks could be used for better weight transfer. See the chassis tuning chapter for more details on using these shocks.

Generally front end settings don't change a lot between track conditions, but camber will be adjusted more than other settings because of the banking of the race track. A greater banking angle generally requires more negative camber at the right front. Check tire temperatures and tire wear to determine required camber changes. Evaluate the differences in tire temperatures across the face of each tire, comparing the inside, center and outside edges of the tires.

Shock absorber travel on the right side of the car will give you a good guideline for proper chassis performance. The shock travel would be the same for a wet, muddy track, a moist hooked-up track and a dry slick track.

Different classes of cars will have base line settings similar to those of the classes of cars shown here, but probably will require some different settings and changes. Cars that are heavier and use track tires will have some set up procedures that vary. For instance, a heavier weight car using a track tire may need a higher rear weight percentage to make it handle properly.

Chapter
15

Track Tuning And Adjustment

Dirt track racing is notorious for changing track conditions. The track will continually change its characteristics from when you first get to the track, into hot laps, heats and features. This means that the driver or crew chief will have to know when and how to make adjustments to the car to correct handling problems on a wet track, tacky track, hooked up track and dry slick track.

First you must determine what kind of track conditions you will be running on. Through the hot lap session, you can learn what the car needs. One factor that you must take into consideration is that the chassis must be set properly before you can make any other adjustments. Camber, caster, toe-out, tire pressure, tire compound, and brake bias must be set properly. If these areas are not set correctly, then you will never stand a chance at setting up your car to track conditions.

Different Types Of Track Conditions

Through an evening of racing at a typical dirt track, a racer will see several different types of track conditions. Generally, a track will start out very wet. The dirt may be very heavy, if the water is worked down into the track, or it may be very slippery if the water is just setting on the very top of a harder dirt surface. Under these conditions, it is very difficult to make a car turn left. The rear wheels are providing forward traction while the front wheels want to skate forward as they are turned.

Once the water is packed down into the track, it produces a tacky track condition. This is a type of surface where the tires can really bite into the dirt

When the track is very wet and slippery, it is very difficult to make a car turn left.

and gain good traction — good side bite and good forward bite.

Finally, as the water evaporates out of the dirt and the dirt packs down, it gets very hard and slick. When the track gets that way, the back end of the car gets very loose and it is difficult to get any forward traction.

Let's discuss some of the different types of track conditions and different handling problems associated with these conditions.

Wet Track Condition

Most probably your car will have an understeer or push condition on a wet track. When we are talking about a wet track, we mean one that has not yet been packed down all the way and has a lot of water still on the surface. This type of condition usually won't stay long and therefore you probably won't be racing

On a wet track there is limited traction. The chassis set-up needs to force the tires to dig into the track surface.

long on this kind of track. Normally this type of track has been rained on by the water truck and takes a little while for it to be packed in. There are adjustments that can be made to help cure problems in this area, but don't overdo it because the track won't stay this way long.

The first thing you must do when the car goes out on the track is make sure the brakes are working properly. This is one of the areas that affects the handling of a race car on any track condition, so double check this every time. The car has a balance bar on the brake pedal assembly and you should make sure that it is working properly. And, make sure you know which way to turn the balance adjuster crank to gain more front brake or more rear brake.

A good rule that many racers go by for balance bar adjustment is: turning the adjuster clockwise puts front brake into the car. Turning it counter clockwise puts rear brake into the car. This is makes the adjustment easy to remember. Make sure you know what your brakes are doing when you turn the adjuster crank.

With a push condition on a wet track, the first thing to check is the brake balance. For this type of track, it should be adjusted for about 75% rear brake bias. This will make the car set or tail the rear out when you are entering the corner. This is the most important factor to get a car to enter the corner properly on a wet track.

This works just the opposite if a car has a loose condition or oversteer on wet dirt. If the problem is oversteer, put some brake bias back into the front.

On a wet, heavy track, it is very easy to lock up the right front wheel, so a brake shut-off valve is used to control this. If the car enters the corner and pushes, make sure the valve is set to shut off pressure to the right front brake. In addition to preventing right front wheel lock up, this also creates a pulling effect to the left when the brakes are applied. This will help the car set up for a corner on a wet track.

Once the car has entered the corner and you have the car set properly, the next step is to make sure it gets around the corner. This sometimes is as difficult as anything else on a wet track because there is limited traction and you are basically sliding around. The chassis set-up needs to force the tires to dig into the track surface, so you need to create side bite and forward bite.

Use as soft a tire as possible, and one that is grooved well for this type of condition. Use plenty of tire stagger in the rear — in the range of 5 to 6 inches. Don't use a lot of left side weight because this usually creates a push condition on a wet track. Move ballast to the right.

Lower the Panhard bar height in order to create more roll and right rear side bite. The ballast should be mounted high in the chassis.

Using a shorter length torque arm (adjusting it to a shorter distance from the rear end to lift point) will usually assist in this type of condition. A torque arm that is too long sometimes will lift the car's front wheels (or lighten up the front end) and create a push condition.

Remember that any time you make an adjustment to correct a push condition, the opposite type of adjustment usually corrects a condition for a loose race car.

The Tacky Track

The wet track condition usually changes quickly to a tacky track, and it usually lasts longer than the extremely wet track. Therefore you should take this type of condition more seriously than a wet track because you will be running on it longer.

Tacky tracks usually cause a car to push. This is because the rear tires grip the track so well. The most important adjustment for a tacky track condition is to free the car up. This means that the chassis has to be loosened up to decrease right rear side bite.

A tacky race track provides a lot of traction. Side bite or body roll is not nearly as critical as it is on a dry track. Photo by Dennis Mattish

Dry track conditions usually require softer springs and shocks, higher ballast placement in the chassis to induce roll, and less left side weight percentage. Photo by Earl Garretson

A tacky race track usually starts sometime during time trials or early in the heat races and usually fades to a dry or rubbered down race track by feature time. So qualifying in time trials or heat races is very critical. This requires a good set up for a tacky track. Most of the time a tacky race track provides a lot of traction. You want a race car that will turn and be extremely fast through the corners due to the traction that will be available in the track. On a tacky track, side bite or body roll is not nearly as critical as it is on a dry track. There are a number of ways the chassis can be adjusted for this condition.

If you are using a Panhard bar in the rear suspension, you need to have the bar set higher in the chassis than normal, and close to level. This raises the rear roll center. This keeps the right rear from planting down hard and causing a push condition in the middle of the corner.

Rear brake bias is another factor. You will need more rear brake than normal on a tacky track to get the car to set, but not as much as used on a real wet track. Stiffer rear springs also help free a car up on this type of track.

A stiffer right rear and right front shock valving helps on a tacky track. These shocks stabilize the car and help weight transfer from occurring too quickly. Change from a 75 to a 76 at the right front, and from a 94 to a 95 at the right rear.

Ballast placement is another chassis tuning tool. Placing the ballast lower on the left side frees the car and helps it turn on a heavy tacky track.

A tacky track will let you take full advantage of a 4-link rear suspension. Creating more rear roll steer and/or increasing the right side wheelbase will help the car turn on tacky tracks.

Keep the torque arm length set at the shortest adjustment to prevent the front end from pushing. A tacky track also requires a lot of stagger — in the 5 to 7-inch range. The tire compound should still be the softest available because a tacky track requires maximum front and rear bite. Brake pressure to the right front brake should still be shut off.

Dry Track Conditions

Dry track conditions usually require softer springs and shocks, higher ballast placement in the chassis to induce roll, and less left side weight percentage. Lowering the Panhard bar will help create side bite and roll on a dry track because it lowers the rear roll center.

Dry tracks usually require a somewhat harder tire compound. Tire compound is definitely an important chassis tuning tool. Always lean toward a harder tire compound if there is any doubt.

Some general chassis tuning guidelines for dry, slick tracks are as follows:

If a car is too tight, soften the front springs and/or stiffen the rear springs.

If a car is loose on turn entry and through the middle of a turn, the left and right front springs should be equal in rate, or add right rear weight.

If a car has four wheel drift in a corner, it has too much low left side weight. Create more side bite by moving the left side ballast higher in the chassis.

If a car is tight on turn entry, soften the left front spring or add left rear weight.

If a car has four wheel drift in a corner, it has too much low left side weight. Create more side bite by moving the left side ballast higher in the chassis. Also, you can lower the rear roll center by lowering the Panhard bar mounting height.

Stagger should be decreased on a dry track. If you used 5 to 7 inches of stagger on a tacky track, decrease it to the 4 to 6-inch range for a dry track.

The right front brake should be operational. The torque arm length should be set at 38 or 40 inches.

Tuning With Wheel Offset

Using a wheel spacer or changing wheel offset makes the car loose or tight, depending on where the change is made. Going from a 5-inch offset right rear wheel to a 3-inch offset wheel (moving the tire out away from the body) will free the car up or make it turn by doing two things:

First, it will place more static weight load on the left rear. Usually a 2-inch offset change will move about 22 to 25 pounds to the opposite side of the car when you move the wheel out. This frees the car at turn entry. Second, with the wheel being further out on the right rear, it will take away side bite and not pin the right rear tire as hard so the car is free all the way around the corner. This helps the car turn easier on a tacky hooked up track.

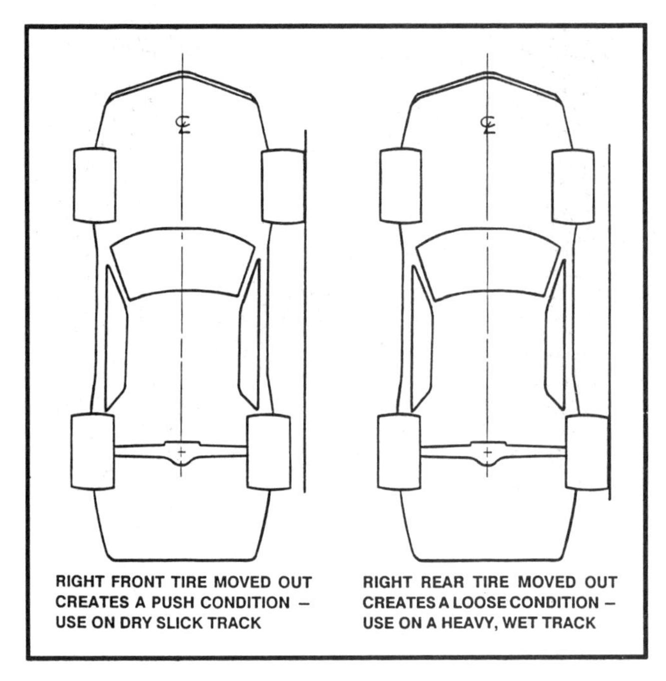

RIGHT FRONT TIRE MOVED OUT CREATES A PUSH CONDITION – USE ON DRY SLICK TRACK

RIGHT REAR TIRE MOVED OUT CREATES A LOOSE CONDITION – USE ON A HEAVY, WET TRACK

The opposite occurs when you space the left rear wheel out away from the car. That tightens the car at turn entry and helps create side bite all the way around the corner. The reason this happens is because spacing the left rear out loads the right rear corner. This tightens the chassis at corner entry, which pins the right rear harder and creates side bite. Using more wheel offset at the right rear and less at the left rear would be a desirable chassis adjustment for a dry slick track.

Tuning With Shock Absorbers

Shock absorbers are extremely important to a car's handling on a dirt track. Shock absorbers control the rate of weight transfer during cornering, they control spring movement, and they control suspension movement over bumps and surface undulations. Being able to control the chassis with the proper shock absorbers is a key element to proper handling on dirt. Shocks can be used to help control handling problems or to induce desirable handling characteristics.

Shocks have no effect on the *amount* of weight that is transferred dynamically during braking, acceleration and cornering. They can, however, affect the transient response in the pitch and roll axis. The amount of weight transferred is dependent on the center of gravity, roll axis, and roll rates. *How*

The slicker or slower a track is, the softer the shock absorber you should use. The faster or heavier a track is, the stiffer the shock should be.

quickly the weight is transferred is controlled by the shock absorbers.

Track speed and the amount of traction available are important elements for choosing the proper shock absorber. For example, the slicker or slower a track is, the softer the shock absorber you should use. The faster or heavier a track is, the stiffer the shock should be. Shock absorbers have to be changed on the car as track conditions change. This keeps the suspension tuned to the track.

If you are in doubt about the choice between two different shock rates, choose the softer one. In almost all cases, softer shocks will yield faster lap times. Softer shocks generally make a car faster because they let weight transfer quicker to a wheel. This puts the transferred weight to work quicker and plants that corner of the car for maximum traction.

Split Valving Shocks

There are conditions when you want to use two different shocks all rolled into one. For example, you may want to keep a particular corner of a car from transferring a lot of weight when it rises up, which would require a stiff rebound valving shock, but you don't want to have a stiff shock for compression or bump travel at that corner. In that case, a split-valving shock can be used. A split valving shock has one rate in compression, and another rate in rebound. A shock that is stiff on rebound valving and softer on compression valving is called a "tie down" shock. It keeps the corner of the car where it is attached tied down, making it hard for the body to rise up at that corner.

With a 75/3 shock on the right front, under acceleration the "3" rebound valving will transfer weight quickly from the right front to the left rear, which tightens up the chassis. Photo by Dennish Mattish

The basic area where we use split valve shocks is when a car needs to be looser on turn entry and tighter at corner exit. If a track is smooth, use a 75/3 shock on the right front. This is a 7-inch stroke shock with a 5 valving in compression and a 3 valving in rebound. With the "3" rebound valving, under acceleration the car will transfer weight quickly from the right front to the left rear, which tightens up the chassis. If the track has banking or is rough, use a 76/3 shock (6 valving in compression, 3 valving in rebound). The "6" valving on compression would help to better control track roughness, or more downforce caused by track banking. These shocks keep weight longer on left rear during corner entry and then transfer weight quicker at corner exit.

Occasionally we will use a 93/5 shock on the left rear (3 valving in compression, 5 valving in rebound). This is a tie down shock. This shock keeps left rear weight in the chassis longer on turn entry and through the middle of the corner. Then it lets weight transfer back to the left rear quicker under compression.

Chassis Tuning With Gear Ratio

Gear ratio is also a factor in tuning a chassis. Our theory on gear ratio is to gear a car as if you could use every single horsepower developed by the engine, and your car was hooked to the max. Remember that the track does not change in size, just condition. Using a slightly lower gear ratio enables the driver to be smoother because he won't bog the engine. If the track becomes hard and slick, the driver can gradually ease into the throttle. The gear should help slow the car down on corner entry and therefore braking won't be as extreme. This in turn makes a more consistent and smoother driver, which creates faster lap times.

On a dry slick track, the gear ratio may need to be changed, depending on where the driver wants to run on the track. If he wants to stay on the very bottom, a lower ratio would be used. If he feels that he can be faster by running at the top of the track, a slightly higher gear ratio would be used.

Chassis Sorting Philosophy

When sorting out handling problems with your car, make sure you are honest with yourself on what the car is doing. Make sure you know where the chassis baseline is at — don't guess. Have all of your chassis settings written down.

If you think the car is loose, really think about if you are having to make the car break loose to make it turn. I've often seen it where a driver has to make his car turn and creates his own loose situation. Yet he didn't realize why his car was actually loose. The real problem is that the car was tending toward understeer going into the corners. Make sure that you are adjusting the chassis for the correct problem.

And remember that all chassis tuning will be in vein unless tire stagger, tire pressure, front end alignment, tire compound, springs, and ride heights are correct. Before any track tuning or adjusting can be done you must have these properly set.

One thing that we have noticed through many years of racing is that a race car which tends toward being loose is faster under all track conditions. When a car is too tight and pushes, the driver cannot be smooth and consistent. Any time a driver is smooth, he is faster. If a driver learns to drive a car in a loose condition, he can adjust to different track conditions much easier.

Another significant chassis adjustment tool is the rear spoiler angle. The range of spoiler angle used is between 20 and 55 degrees. On a shorter or slicker track, more spoiler angle is used. A longer, hooked up track requires less angle. A slick stop-and-go track might require 50 degrees of spoiler angle, whereas a 3/8-mile track with good moisture may require 40 degrees. A very fast 1/2-mile track with long sweeping turns may only require 20 degrees. If the car starts to push, take spoiler angle out.

What you have to realize when driving a race car is that just because a car feels faster to you does not mean that it *is* faster. Time and time again I have seen lap times be so much quicker when the driver makes smooth consistent laps. Keep the car under you and let the car roll through the corners. Remember that once you get a car hung out with oversteer, you are losing time.

Chapter
16

Race Car Maintenance

Race car maintenance has won or lost more races than any other area of racing. Neglecting maintenance is a major reason that wins don't add up. Even a team on the smallest budget can upgrade their program if maintenance is done properly.

The first thing you have to do is design a good program for maintenance. Have a set-up sheet or a guideline sheet to use for maintenance procedures. Having this sheet enables you to check off each item as it is completed so you don't forget any area of scheduled maintenance. Once you have this list, you can go over it every week and keep records of all parts checked and/or replaced. One of the nicest maintenance sheets I've seen used is the maintenance sheet used by GVS Racing and Freddy Smith, which is reproduced in this chapter.

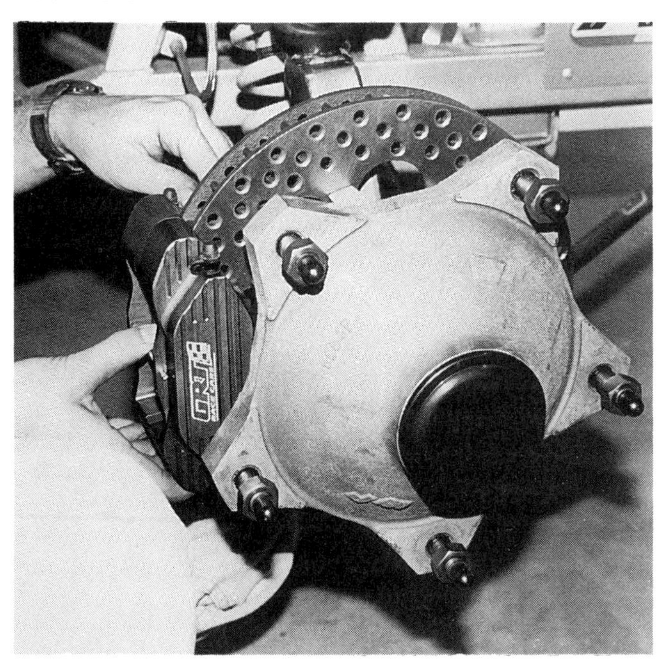

Visually inspect the brakes, rotors and all suspension components.

A good maintenance sheet will remind you to check the following:

Front End Parts

Ball joints, rod end bearings, rotors, calipers, rack shaft, rack U-joints, hub bearings, shock bearing ends, brake lines and fittings, brake pads, each and every bolt and nut on a front end.

Rear End Components

Trailing arms, rod end bearings, Panhard bar and end bearings, torque arm and braces, bird cages, caliper brackets, calipers, brake pads, brake lines and fittings, rotors, hub bearings, drive flanges, axles, yoke, drive shaft U-joint and bolts, shock bearing ends on rear end and fifth coil, quick change gears and lower shaft, and pinion. Check each and every nut and bolt on all rear end components.

Fluid Levels And Lubrication

Lubricate rod end bearings on trailing arms, tie rods, struts, shocks, steering, brake adjusters, quick release steering hubs, shifting linkage, carb linkage, lug studs, caliper pistons and any other moving part that does not have a grease fitting. Use a good spray lubricant so the lube penetrates. Grease all ball joints, rack shafts, bird cages, and drive shaft U-joints and any part with a grease fitting. Check all fluid levels in the rear end, transmission, dry sump tank, power steering tank, radiator and brake fluid.

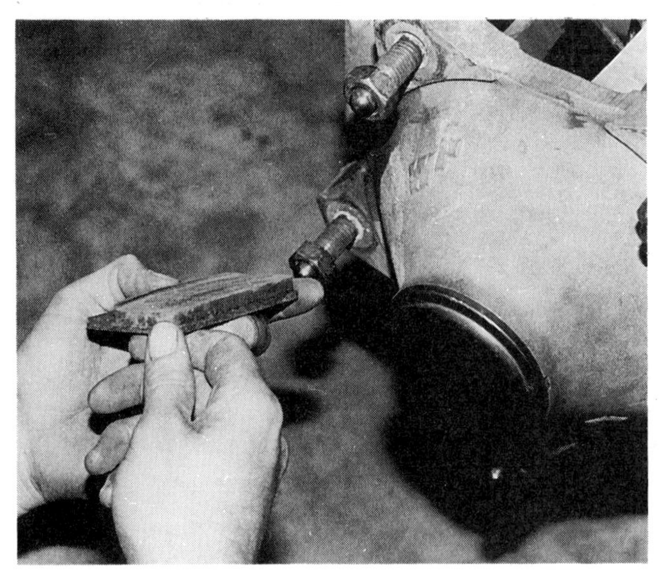

Inspect brake pads for wear. Measure pad thickness each week so you have an idea of wear rate.

Use a good quality spray lubricant on all bearings. The lubricant should be penetrating and not evaporate.

Check all fuel lines and fuel filters. Check all belt tensions and conditions.

Engine Maintenance

Each week check: valve springs, valve lash, timing, float levels, spark plug gaps and conditions, oil and Oberg filters, air cleaner and engine breathers, and nuts and bolts in the entire engine area.

Fluid Changes

Oil should be changed on a regular basis depending on the type of oil used and what your engine builder recommends. Oberg filters give you a idea of what is going on inside an engine and should be checked after each night of racing.

Brake fluid should be changed each month. Transmission fluid should be changed 3 times a season. Shocks should be taken off the car and checked at least once a week. Send shocks back to the manufacturer for a check up once a year.

Nuts, Bolts & Other Components

Keep an eye on highly used nuts and bolts. Sometimes threads will become stripped and should be repaired immediately with a Helicoil or by retapping. Keep anti-seize on these type of fasteners.

Bolts should be torqued and checked periodically. Any bolts or nuts that are not removed or installed frequently should be safety wired if Nylocks are not used.

Maintenance goes further than just your race car. Keep a check on all spare parts and components. Check wheels, bead locks, or any other item that you might be bolting to or using on your race car. Take the time to do it right. You'll be proud you did.

Don't ever underestimate the power of good maintenance. You cannot win if you do not finish.

Grease the bird cages, and everything else with a grease fitting, every week.

G V S RACING, INC.
WEEKLY & BI-WEEKLY MAINTENANCE PROGRAM

DATE ____________

CAR ____________ MOTOR ____________

1 ___ ☐ CHECK WATER
2 ___ ☐ CHECK POWER STEERING FLUID
3 ___ ☐ CHECK CLUTCH FLUID
4 ___ ☐ CHECK BRAKE FLUID
5 ___ ☐ CHECK TRANSMISSION FLUID
6 ___ ☐ CHECK REAR-END FLUID
7 ___ ☐ CHANGE OIL
8 ___ ☐ CHANGE OIL FILTER & INSPECT
9 ___ ☐ CLEAN AIR FILTERS & VALVE COVER BREATHERS
10 ___ ☐ CHECK BRAKE PADS
11 ___ ☐ BLEED BRAKES
12 ___ ☐ CHECK HUBS & ROTORS
13 ___ ☐ CHECK DRIVE FLANGES & LUBRICATE
14 ___ ☐ PACK WHEEL BEARINGS
15 ___ ☐ CHECK & SET TOE-END
16 ___ ☐ CHECK & SET CASTER/CAMBER
17 ___ ☐ CHECK FRONT SUSPENSION
18 ___ ☐ CHECK REAR SUSPENSION
19 ___ ☐ CHECK LIFT BAR
20 ___ ☐ CHECK RAP SHOCK BRKTS.
21 ___ ☐ CHECK SHOCKS
22 ___ ☐ CHECK SPRING
23 ___ ☐ CHECK NUTS & BOLTS
24 ___ ☐ STRAIGHTEN/REPLACE BODY COMPONENTS

25 ___ ☐ CLEAN & GREASE SHACKLES
26 ___ ☐ STRAIGHTEN/REPLACE BUMPERS, BRACES, DOOR BARS
27 ___ ☐ CHECK FENDER WELL CLEARANCE
28 ___ ☐ CHARGE BATTERY
29 ___ ☐ GREASE

ENGINE RELATED:

30 ___ ☐ CHECK VALVES (358) WEEKLY, CHECK ROCKER ARMS
31 ___ ☐ CHECK BELTS, LINES, BRACKETS, LINKAGES
32 ___ ☐ CHECK SPARK PLUGS & LUBRICATE
33 ___ ☐ CLEAN & LUBRICATE BY-PASS VALVE
34 ___ ☐ SET FUEL PRESSURE
35 ___ ☐ FUEL/ALCOHOL/RACING GAS
36 ___ ☐ CHANGE FUEL FILTER
37 ___ ☐ DRAIN WATER FROM BREATHER LINE
38 ___ ☐ PUT LEAD IN OR OUT
39 ___ ☐ SCALE RACE CAR
40 ___ ☐ SET-UP

LET'S RACE ... !

VERY IMPORTANT !!!
DRAIN ALCOHOL CARB BOWLS WEEKLY AND FILL WITH GAS.

DO NOT LET ALKYL CARBS SIT ALONE UNLESS FILLED WITH GAS AND OPERATING SECONDARIES, BUTTERFLY FOR LUBRICATION.

REMARKS: ____________

Chapter

17

Repairing A Crashed Car

Determining The Extent Of Damage

When you crash a car, damage may be extreme, or it may be minimal. I've seen race cars that are crashed extremely hard — sheetmetal body parts and bent parts all over. Your first thought: is it totaled out? Can it be fixed?

Sometimes cars appear that way, but the first thing to do before making any assumptions is remove all of the broken and torn up panels and parts. Get everything out of the way so you can determine the damage. A lot of times, once you view the chassis, you can tell it is mostly cosmetic damage. For instance, the way GRT cars are built with crash zones, the damage usually doesn't get into the critical points or suspension mounts. If you just remove a tail section or take the radiator out, then you can repair the damage and the car will look as if it has never been crashed.

If you crashed your car, do you have a set of written reference points so you can tell what is bent?

The main thing to look for is suspension part damage. These are the areas that are most critical. When you get your new race car, look it over well. Notice how snout bars, tail section bars, and main frame rails are constructed. These are areas that you can look at and tell if something is bent. Take measurements of suspension parts to reference points, and keep records of this. That way you can go back and check the measurements in case of a crash.

One way to tell if a race car has been bent is to see if the car sets on jack stands differently than it normally does. Always work on your car in the same area of the floor. Mark four spots on the shop floor where your jack stands set. Notice how the main frame of the car sets down on the jack stands. If, for any reason after a crash, the car rocks back and forth or has more or less clearance than before, there is a good chance the chassis is twisted or bent.

If you are the least bit concerned about your car, consult your chassis builder for assistance. If the car has been bent somewhere in the suspension area, then most likely you will need to have your car rejigged or checked to ensure the chassis is not damaged.

What I suggest is to take any prior measurements and recheck them. If any measurement is off more than .0625-inch or so, start looking for the damaged area. If you are not sure about damage, call your chassis builder and get more reference points from the jig.

Sometimes the damage is hidden in the frame or chassis. We have cars that have turned over in a flip and they don't appear to be bent, but when you put them in the jig, they won't fit. They are twisted pretty bad.

Frame rail angles taken with a machinist's protractor at specific places can reveal damage to the chassis not measurable from reference points.

On the other hand, a car that has run into a wall and then been hit by another car may just need some bumpers and minor frame repair. Always remember in racing there is no way to tell what is going to happen or how bad it is going to be. Don't take anything for granted, and don't be surprised if you can't find anything totally destroyed after a crash.

If a race car chassis is bent too badly and you have many repairs to perform on it, sometimes it is better to replace the chassis. Chassis are about the cheapest part of the car. If the cost of repairs is going to be close to, or as much as, a new chassis, then get a new one.

Looks can be deceiving. Although it appeared that there was major damage here, careful measurement revealed no structural movement.

Only measuring from all of the front reference points can tell you where the real damage is with a wreck like this.

Should we scrap it or rebuild it? Only careful checking, and comparing measurements and angles to your notes, will give you the complete answer.

Suspension Component Damage

Suspension component damage is an area of major concern for any racer. Any component that has been involved in or near a crash probably will be damaged in some way. Take everything apart and visually inspect all parts thoroughly. If you get hit hard in a rear wheel, most of the time there is a bent rear end tube. Or, if it is an aluminum tube, it will be broken at the snout. I've seen many aluminum rear ends be involved in a crash and then they are put back together without checking the axle tube. The next thing you know, the brake caliper is the only thing that is holding the wheel on. So inspect for this if you use an aluminum tube rear end.

Any suspension component that has been involved in or near a crash probably will be damaged in some way. Take everything apart and visually inspect all parts thoroughly. This wheel is bent, so suspect damage to all of the bolt-on components behind it, including the entire steering system.

Most of the time, steel tube rear ends bend. You can take some quick measurements from hub to hub, or sidebell to hub, to check for squareness. A lot of times, the axle will be difficult to remove, and this is another sign of a bent tube.

Bolt-on components like rod ends, aluminum tie rod tubes, shocks, bird cages, calipers, rotors, hubs, A-arms, spindles, etc., should be thoroughly checked and replaced or repaired if there are any signs of damage.

Shocks can be checked somewhat by hand. To do this, compress the shock all the way in and then pull all the way out. If there are any hard or soft spots, or fluid leakage, have the shock repaired or replace it. A racer should periodically have shocks checked on a dyno for maximum performance and to ensure that the shocks are not bad.

Check rod ends for stress cracks or loose balls. If a rod end has been bent, don't straighten it – replace it. Replace aluminum tie rods tubes also. If it is bent, don't straighten it.

Bird cages sometimes can be straightened if they do not have major damage. Make sure the ears on the bird cage are straight and that the bird cage rotates freely on the axle housing.

Calipers should be visually inspected for leaks or cracks around the mounting points. Rotors should be checked for any cracks on the surface and mounting points. Hubs sometimes will have cracked ears where the wheels bolt, so check that area thoroughly.

A-arms should be checked for twisting or cracking. Usually ball joints will be bent in front end impacts, so replace them.

Spindles are easy to bend in a crash, and they are very critical for proper handling. If a spindle has any damage, replace it! Or, send it back to the manufacturer for repair. No options here.

If you get hit or crash the front end or rear end of a race car, check every single part that is connected to that area of the race car. Most components that are tied in to the area of the crash will be affected now or later, so check it over well.

In racing, the best way to repair is to replace. If you don't, the part will usually fail some time in the future. You could be leading a $3,000 or $10,000 to win race, and then "snap." There goes a $200 component that should have been replaced.

There will be other situations where you can make emergency repairs at the track just to get back on the race track and try to win the race. Make sure in this case that any repairs are done well enough to hold together and not create more damage or jeopardize someone else's car. Racing with a poorly repaired race can be extremely dangerous. Use your head in this area and don't take a chance on something that may not work. There will be other races.

Installing A New Front Clip

Installing a new front clip is a project you can do yourself if you have basic fabrication skills and you have the proper equipment. If you don't have either of these, then let your chassis builder handle the job.

Where do you start? First you must determine how much of the area is damaged, and where to cut off the damaged area. Most of the time a cut is made at the point where the main lower frame meets the part of the clip that rakes upward toward the front. Usually there will be a butt weld at this point that makes a good place to cut. If it is a mandrel bent frame, then the new clip will be welded back together. It would not be practical to replace a whole frame rail just because it was not butt welded before.

After you have cut the lower frame rails, cut the upper snout bars at the point where they tie into the dash bar so they don't have to be sectioned. If you cut the frame in these locations, the job will be performed in a manner that looks original. The reason is that you are welding the frame back together exactly where the original frame was welded together.

Once you have everything cut and the attachment surfaces are ground smoothly, you are ready for the initial placement of the clip. You will probably perform this job on a good flat shop floor unless you own a jig. The next step is to place the clip close to where the old clip went. Support it with jack stands and c-clamps while you are checking measurements.

All this job requires are measurements and squareness. For instance, you will have measurements from a reference point on the rear of the chassis (such as a lower rear suspension mounting point). That will determine where the front clip will be installed from front to back. You can perform this routine on both sides to make sure the clip is square. The side-to-side location will be determined by matching up frame rails at the main frame to the beginning of the new clip. There is only one place it can go from side- to-side. If your front clip doesn't line up from side-to-side, then most probably you need more than a front clip job.

You will now have to determine vertical measurements for leveling the main part of the front clip. Provided the main frame rails are level on the floor, the frame rails of the front clip you are installing will be level also at the point where they start going forward from the upward rake. You can double check this measurement with a measurement from the floor to the top of the front frame housing. This measurement can be taken off of the jig and applied to the shop floor. (We are assuming you are using the floor surface as the jig surface for reference points.)

Once you have determined front-to-back, side-to-side and up-and-down measurements, double check the alignment and all measurements. Tack weld the clip in place, and check one more time before final welding. Now all you have to do is install all of the structural roll cage tubing and weld everything together. Your car will be good as new.

If you have to splice any roll cage tubing, it is best to insert a smaller diameter piece of tubing inside the spliced area. This ties a butt weld joint together. Use a rosette weld, above and below the butt joint, to attach the insert piece to the cage tubing. When the butt joint is welded, you can really penetrate and this actually ties three pieces of tubing together instead of two.

Chapter

18

Record Keeping

Keeping track of exactly what happens, how it happened, track conditions, weather, etc., is very important. In all my years of racing, I have learned that you can't keep enough records. I've seen some racers that don't keep any and I wonder how they can be competitive. They usually aren't. A very competitive racer like Bill Frye studies his records thoroughly. He makes it a point to write everything down after a race while it is fresh in his mind. He has notebooks and notebooks full of every conceivable kind of record that pertains to racing. Every time we return to a track that we have already raced at, everything is there. There is no guessing. I cannot stress enough how much this has been a big part of Bill's success and any other racer that excels.

Basic Record Keeping

First make a basic outline to use and have several copies made. That way you can keep records from night to night and track to track. These are some of the most important items to keep records of.

Race track and date you raced
Weather conditions (this plays a major role in track conditions)
Track conditions (hot laps, time trials, heat and feature)
Track size
Gear ratio (and if you changed and why)
Tire compound and tire size (hot laps, time trials, heat, feature)
What the tires looked like or how they wore after each event
Tire grooving patterns
Engine records (carb jet changes, header changes, etc.)
Fuel usage after each race
Set up records (before you leave the shop)
Corner weights
Spring rates
Shock changes
Spring changes
Air pressure
Front end alignment
Fuel in car
Ballast placement
Weight percentages
Panhard bar placement and length
Radius rod placement and length
Torque arm length
Tire stagger (and any changes made)
Spoiler length
Size of engine

The initial set-up information should be recorded before you leave the shop. Now comes the hard part: writing down everything that is changed and why, what it did for the car and when you changed it. It is best to have someone help with record keeping at the track because the driver is usually occupied with racing and working with the car.

Keep track of lap times at the track, from hot laps to the feature. Not only keep track of your own times, but keep track of the competition also. This enables you to determine when the track is changing and gives you a head start on tire choices and set-up. Keep a good record of tires!

If you did not win the race, or even if you did, keep records of what your competition is doing if you know (tire choices, particulars about the way the car was handling, etc.).

The first thing when you arrive back at the shop, completely rescale the car. Do not change a thing on the car. Scale it and check these numbers against your records that you kept at the track and when you left the shop. This enables you to know exactly what set-up you did have and if any calculation you made was right or wrong.

I cannot stress enough how important records are. The next time you see Bill Frye at a race track, and he is sitting on a tire in the hauler with his head between the pages of a book, you can bet it's not a funny book!

RACING CONDITIONS AND EVALUATIONS

Date________ **Speedway Name** ____________________________ **Location** ________________________

Qualifying Position __________________ **Qualifying Time** ______________________________________

	Starting Position	Finishing Position	No. of Laps	Duration of Race
Heat Race	________	________	________	
Feature Race	________	________	________	________

COMMENTS:

Qualifying ____

____

____

Heat Race____

____

____

Feature Race ____

____

____

	Tires		Shocks		Springs	
Qualifying	LF _______	RF _______	LF _______	RF _______	LF _______	RF _______
	LR _______	RR _______	LR _______	RR _______	LR _______	RR _______
	Stagger_______					
Heat Race	LF _______	RF _______	LF _______	RF _______	LF _______	RF _______
	LR _______	RR _______	LR _______	RR _______	LR _______	RR _______
	Stagger _______					
Feature Race	LF _______	RF _______	LF _______	RF _______	LF _______	RF _______
	LR _______	RR _______	LR _______	RR _______	LR _______	RR _______
	Stagger _______					

Corner Weights

Before	LF _______	RF _______	After	LF _______	RF _______
	LR _______	RR _______		LR _______	RR _______

J-Bar Position ____________________________ **Clamp Bracket** ____________________________

Chapter

19

Suppliers Directory

ACPT
(Advanced Composite Products & Technology Inc.)
15602 Chemical Lane
Huntington Beach, CA 92649
(714) 895-5544 Fax (714) 895-7766
Carbon fiber driveshafts

AFCO Racing Products
(American Fabricating Co.)
P.O. Box 548
Boonville, IN 47601
(812) 897-0900 Fax (812) 897-1757
Coil springs, brake calipers and rotors, shock absorbers, all types of suspension components

Appleton Rack & Pinion
110 Industrial Drive, Bldg. E
Minooka, IL 60447
(815) 467-1175 Fax (815) 467-1179
Power rack and pinion steering systems, suspension components

Russell Baker Racing Engines
505 Henely
Miami, OK 74354
(918) 540-2800
Racing engines

Bert Transmission
395 St-Regis North
St-Constant, PQ J5A2E7
Canada
(450) 638-2960 Fax (450) 638-4098
Racing transmissions

Brinn Inc.
1615 Tech Dr.
Bay City, MI 48706
(517) 686-8920 Fax (517) 686-6520
Racing transmissions

ButlerBuilt Motorsports Equipment
P.O. Box 459
Harrisburg, NC 28075
(704) 784-1027 Fax (704) 784-1024
Seats and dry sump tanks

Carbone Industries
43976 Choptank Ter.
Asdhburn, VA 20147
(703) 858-7822 Fax (703) 858-7823
Carbon/carbon brake rotos

Carrera Shocks
5412 New Peachtree Rd
Atlanta, GA 30341
(770) 451-8811 Fax (770) 451-8086
Shock absorbers, coil-overs, springs

Coleman Machine
N-1597 US 41
Menominee, MI 49858
(906) 863-8945 Fax (906) 863-7027
Driveshafts, brake components, suspension components

D&M Performance Mfg.
P.O. Box 696
Dumas, TX 79029
(800) 233-0571
(806) 935-2448 Fax (806) 935-2998
Racing seats

Deist Safety
641 Sonora Ave
Glendale, CA 91201
(818) 240-7866 Fax (818) 244-9230
Safety equipment, fire systems

Dynatech Headers
P.O. Box 608
Boonville, IN 47601
(800) 848-5850
(812) 897-3600 Fax (812) 897-6264

DynoMax Mufflers
1201 Michigan Blvd.
Racine, WI 53402
(414) 631-6352 Fax (414) 631-6443

Earl's Performance Products
P O Box 10360
Bowling Green, KY 42102
(270) 782-2900 Fax (270) 745-9590
High performance hoses and fittings

Fuel Safe Fuel Cells
Aircraft Rubber Mfg.
63257 Nels Anderson Rd.
Bend, OR 97701
(541) 388-0203 Fax (541) 388-0307
Fuel cells

Fluidyne
4850 E. Airport Drive
Ontario, CA 91761
(800) 266-5645
(909) 390-3944 Fax (909) 390-3955
Oil to water heat exchangers

GRT Race Cars Inc.
83 S. Broadview St.
Greenbrier, AR 72058
(501) 679-2311 Fax (501) 679-5496
Dirt late model and IMCA modified chassis builder

Hawk Brake
920 Lake Road
Medina, OH 44256
(800) 542-0972 Fax (330) 725-1643
Brake pads

Howe Enterprises
3195 Lyle Rd
Beaverton, MI 48612
(517) 435-7080 Fax (517) 435-3331
Suspension components

JET-HOT Coatings
55 E. Front St.
Bridgeport, PA 19405
(800) 432-3379 (610) 277-5646 Fax (610) 277-5736
Specialty metal coatings

KSE Racing Products
P.O. Box 821
White House, TN 37188
(615) 672-5117 Fax (615) 672-2366

Kilsby Roberts Co.
5207 Scott Hamilton Dr.
Little Rock, AR 72209
(501) 568-4371 Fax (501) 568-7869
(800) 876-3116 National office
(Offices in most metropolitan U.S. areas)
Structural tubing

Marsh's Racing Tires (MRT)
22624 Marsh Rd.
Siloam Springs, AR 72761
(501) 524-4157 Fax (501) 524-5168
Racing wheels, racing tires

Mittler Bros. Machine & Tool
P O Box 110
Foristell, MO 63348
(800) 467-2464
(636) 463-2464 Fax (636) 463-2874
Metal fabrication equipment

Oberg Motorsports Inc.
6120 195th St. NE
Arlington, WA 98223
(360) 435-9100 Fax (360) 435-5598
Filtering systems

Outlaw Brakes
Reb-Co Racing Enterprises
5130 Piney Grove Rd.
Cumming, GA 30130
(770) 844-1777
Disc brakes

Performance Friction Corp.
83 Carbon Metallic Hwy
Clover, SC 29710
(803) 222-2141 Fax (803) 222-2144
Brake pads

PRO Shocks
1715 Lakes Parkway
Lawrenceville, GA 30043
(770)995-3600 Fax (770) 513-4406
Shock absorbers, coil-overs, springs

Professional Racers Emporium (PRE)
2376 Main St.
Riverside, CA 92501
(909) 779-1300
Race car fabrication, suspension components

Quarter Master Industries
510 Telser Rd
Lake Zurich, IL 60047
(847) 540-8999 Fax (847) 540-0526
Clutches, hydraulic release bearings

RAM Automotive Co.
201 Business Park Blvd.
Columbia, SC 29203
(803) 788-6034 Fax (803) 736-8649
Clutches, hydraulic release bearings

RCI
12440 Hwy 155 S.
Tyler, TX 75703
(903) 581-5976 Fax (903) 581-1108
Fuel cells and safety equipment

Royal Purple Synthetic Lubricants
23763 S. Owens Rd.
Porter, TX 77365
(281) 354-7788 Fax (281) 354-7335

RTC
(Racing Transmission Components)
1000 E. 7th St.
St. Paul, MN 55106
(612) 776-9781 Fax (612) 778-1166
Shifter for racing Powerglide transmissions

Schoenfeld Headers
20 Cane Hill St.
Van Buren, AR 72956
(501) 474-7529 Fax (501) 474-2249

Sierra Racing Products/ Dan Press Industries
1397 State Highway 506
Vader, WA 98593
(360) 295-3436 Fax (360) 295-3486
Disc brake systems, Gold Trak differential

Speedway Engineering
13040 Bradley Ave.
Sylmar, CA 91342
(818) 362-5865 Fax (818) 362-5608
Suspension components, quick change rear ends

Speedway Motors
P O Box 81906
Lincoln, NE 68501
(402) 323-2101 Fax (800) 736-3733
They sell almost every part necessary for building a race car. Their catalog is a must.

Sweet Manufacturing, Inc.
3421 S. Burdick
Kalamazoo, MI 49001
(616) 344-2086 Fax (616) 384-2261
Rack and pinion steering, power steering, spindles, steering u-joints, suspension brackets

TCI Automotive
One TCI Drive
Ashland, MS 38603
(601) 224-8972 Fax (601) 224-8255
Race prepared Powerglide transmissions

Tilton Engineering
P.O. Box 1787
Buellton, CA 93427
(805) 688-2353 Fax (805) 688-2745
Pedal assemblies, master cylinders, brake proportioning valves, hydraulic clutch bearings, flywheels, starters

Weld Racing
933 Mulberry Street
Kansas City, MO 64101
(816) 421-8040 Fax (816) 283-3430
Racing wheels

Wilwood Racing Products
461 Calle Bolero
Camarillo, CA 93012
(805) 388-1188 Fax (805) 388-4938
Brake systems, hubs, rack and pinion steering

Winters Performance Products
2819 Carlise Rd.
York, PA 17404
(717) 764-9844 Fax (717) 764-0617
Quick change rear ends, hubs

Woodward Machine Corp.
P.O. Box 4479
Casper, WY 82604
(888) STEER-US
Power rack and pinion steering systems

Wrisco Industries Inc.
355 Hiatt Dr., Ste. B
Palm Beach Gardens, FL 33418
(561) 626-5700 Fax (561) 627-3574
(800) 627-8848
Prefinished aluminum panels

STEVE SMITH AUTOSPORTS

The Racer's One-Stop TECH Source

The Complete Karting Guide
For every class from beginner to enduro. Covers: Buying a kart/equipment, Setting up a kart — tires, weight distrib., tire/plug readings, exhaust tuning, gearing, aerodynamics, engine care, competition tips. **#S140...$16.95**

4-Cycle Karting Technology
Covers: Engine bldg. & blueprinting, camshafts, carb jetting & fine tuning, fuel systems, ignition, gear ratios, aerodynamics, handling & adjustment, front end alignment, tire stagger, competition driving tips. **#S163...$16.95**

Kart Racing – Chassis Setup
How a kart chassis works & how to improve it. Covers: steering principles, weight distrib., tire theory & use, chassis setup & alignment, trackside chassis tuning, & more. For 2 & 4-cycles, asphalt & dirt. **#S245...$16.95**

Kart Driving Techniques
By Jim Hall II. A world-class driver and instructor teaches you the fundamentals of going fast in any type of kart. Includes: kart cornering dynamics, learning the proper line, learning braking finesse, how to master trail braking, racing & passing strategies, getting out of trouble, & much more. For all drivers from novice to expert. Includes a chapter on shifter karts. **#S274...$16.95**

Beginner's Package

1) Building A Street Stock Step-By-Step
A complete guide for entry level stock class, from buying a car and roll cage kit to mounting the cage, stripping the car, prep tips, and mounting linkages, etc. **#S144...$15.95**

2) The Stock Car Racing Chassis
A basic stock car handling book. Covers: Under and oversteer, The ideal chassis, Spring rates and principles, Wedge, Camber/caster/toe, Guide to bldg. a Chevelle/Monte Carlo chassis. **#S101...$10.95**

3) So You Want To Go Racing?
Everything you need to know to get started in racing. **#S142...$11.95**
Special Offer: Buy all 3 books above together and the special package price is only **$33.50.** Shipping charge is $10.

The Racer's Math Handbook
An easily understood guide. Explains: Use of basic formulas, transmission & rear end math, engine, chassis & handling formulas, etc. Includes computer program with every formula in the book & more. **#S193C..$24.95**

The NEW Racer's Tax Guide
How to LEGALLY subtract your racing costs from your income tax by running your operation as a business. Step-by-step, how to do everything correctly. An alternate form of funding your racing. **#S217..$17.95**

Advanced Race Car Suspension
The latest tech info about race car chassis design, set-up and development. Weight transfer, suspension and steering geometry, calculating spring/shock rates, vehicle balance, chassis rigidity, banked track corrections, skid pad testing. **#S105.. $16.95** WorkBook for above book: **#WB5..$10.95**

Racer's Guide To Fabricating Shop Eqpment
An engine stand, hydraulic press, engine hoist, sheet metal brake and motorized flame cutter all for under $400. Step-by-step instructions. Concise photos and drawings show you how to easily build them yourself. **#S145..$16.95**

Short Track Driving Techniques
By Butch Miller, ASA champ. Includes Basic competition driving, Tips for driving traffic, Developing a smooth & consistent style, Defensive driving tactics, and more. For new and experienced drivers alike. **#S165...$16.95**

Bldg. The Pro Stock/Late Model Sportsman
Uses 1970-81 Camaro front stub with fabricated perimeter frame. How to design ideal roll centers & camber change, fabricating the car. Flat & banked track setups, rear suspensions for dirt & asphalt, track tuning. **#S157...$21.95** Special package price for book & Bldg. The Stock Stub Race Car video is only $55.95

Sprint Car Chassis Set-Up
Detailed info on chassis set-up and alignment • Birdcage timing • Front end geometry • Choosing torsion bar rates • Chassis tuning with shocks • Using stagger and tracking • Tire selection, grooving, siping, reading • Blocking, tilt, rake, weight distrib. • Track tuning • Wings. **#S174 ...$19.95** Special Package Price for Sprint Car Chassis Setup Video & book is $53.95.

Mini Sprint/Micro Sprint Racing Technology
Covers frames, suspension, chassis adjustment & setup, blocking, wt. dist., chain drive systems, chassis squaring, tires, wheels, wings, & much more. **#S167...$17.95**

How To Build Chev Small Block V8 Race Engines
For the ultimate in affordable performance. Details every step of building: tear down, machine work, porting & cylinder head work, assembly, break-in, tuning & inspection. Affordable high perf. tips & techniques. **#S211...$19.95**

Hot Holleys
Extensive tuning & modification info on the Holley 4150HP & Dominator. Covers all CFM sizes of 4150HP. Shows complete gas & alcohol buildup of a 390 CFM late model/modified carb. Complete alcohol tech. **#S278..$19.95**

Chevy Small Block Part & Casting # Guide
New 3rd edition. Covers all small block heads, blocks, intake & exhaust manfolds from 1955 – 1999. **#S208..$15.00**

Trailers — How To Design & Build
All the tricks of the trade for building a race car trailer. Details on suspension, materials, construction, braking systems, couplers. **#S203...$24.95**
Trailer Building Blueprint #B800...$17.95 Special Package Price for blueprint & S203 book is only $36.95.

Trackside Chassis Tuning Guide
A handy, concise, indexed guide to analyzing and correcting race car handling problems. Organized by problem (such as corner entry understeer, etc.) with proven solutions. Almost every solution to every chassis problem is contained. 3.5"x7.5" **#S283...$15.95**

How To Run A Successful Race Team
Covers key topics such as team organization and structure, budgeting & finance, mngmt., team image, goal setting, scheduling & more. A valuable asset for all racers. **#S265...$17.95**

The Great Money Hunt
This book is ONLY for those who are really SERIOUS about obtaining sponsorship. Provides strategy with proposals, justifications & sample correspondence. How to find prospects, & sell and fulfill sponsor's marketing needs. Detailed & comprehensive. **#S200...$69.95**

Get Sponsored
By Charlie Hayes How to create a sponsorship program & proposal with a 7-step approach. A guide to finding prospects, sample correspondence, contract agreement. **#S260...$59.95**

Nuts, Bolts, Fasteners & Plumbing Handbk
How to choose and use hardware & plumbing • Rivets & quick release fasteners • How to engineer joints, brackets & mounts • What each nut, bolt, fastener & plumbing piece is, how it works, how to torque it, where to use it. **#S177...$19.95**

Sheet Metal Handbook
How to form aluminum, steel & stainless. Includes: equipment, bldg. interiors, layout, design & pattern-making, metal shaping, & more. **#S190...$17.95**

Welder's Handbook
Weld like a pro! Covers gas, arc, MIG, TIG, & plasma-arc welding & cutting. Weld everything from mild steel to chrome moly, aluminum, magnesium & titanium. **#S179...$17.95**

COMPUTER PROGRAMS

Computerized Chassis Set-Up
This WILL set-up your race car! Selecting optimum spring rates, cross weight and weight dist., computing CGH, front roll centers, stagger, ideal ballast location, final gearing & more. Windows 95/98 Format. **#C155... $99.95**

Racing Chassis Analysis
You won't ever need a bump steer gauge again! This program does a complete bump steer analysis with graphs, camber & roll center change. Plus a complete dynamic cornering analysis. **#C176...$99.95**

Engine Shop Computer Program
For most all small & big block Chevy, Ford & MoPar V8s. Build your engine on the computer with cataloged parts. Then it simulates a dyno & plots torque & horsepower output. Then change a part & plot the difference. Very accurate! Windows 95/98 **#C216..$74.95**

TirePro Computer Program
Quickly & easily performs a detailed analysis of tire temps, providing info on: Tire pressure, camber, weight, springs & more. Windows format, on CD-ROM **#C178...$49.95**

Race Car Simulator – 2
Shows effects of chassis changes. Specify chassis & track specs, then the computer shows how the car will handle. Change springs or weight and the computer tells the new handling. Windows 95/98 **#C247...$99.95**

Front Suspension Geometry PRO
Does 3D analysis & graphs all suspension movements including camber curve, caster curve and bump steer. Makes suggestions on what parts to move & where to improve suspension geometry. For all cars, Windows 95/98, on CD-ROM. **#C249...$119.95**

Stock Car Dirt Track Technology
By Steve Smith. Includes: Dirt track chassis set-up & adjustment, rear suspension systems (fifth-coil/torque arm, 4-link, leaf springs), front suspension, shocks, tires, braking, track tuning, construction & prep tips, & much more. **#S196...$24.95**

Street Stock Chassis Technology
By Steve Smith. Includes: Performance handling, chassis & roll cage fabrication, front suspension alignment, changing the roll center & camber curve, rear suspensions, springs & shocks, stagger, gearing, set-ups for asphalt & dirt, wt adjustments, track tuning, & more! **#S192...$24.95** Special Package Price for this book & Street Stock Chassis Setup Video is $58.95

Dwarf Car Technology
By Steve Smith. Detailed chassis & set-up procedures to help you go fast! Front suspension & steering, Rear susp. & driveline, Complete chassis set-up, Track tuning. For both dirt & paved tracks. **#S225...$22.95** Special Package Price for this book & Dwarf Car Chassis Setup Video is $56.95

Expert Chassis Set Up Videos

Chassis Setup With Tire Temperatures	#V232	$19.95
Paved Track IMCA Mod. Chassis Set-Up	#V238	$39.95
IMCA Modified Dirt Chassis Setup	#V184	$39.95
Understanding & Tuning With Shocks	#V266	$39.95
Dirt Late Model Chassis Setup NEW	#V261	$39.95

Aerodynamics For Race Cars
Understanding aerodynamics is necessary to win. Explains: rear spoilers, downforce, front air dams, wings, drag, lift, ground effects, aerodynamics effects on handling. **#S257...$17.95**

Dirt Stock Car Fabrication & Preparation
By Joe Garrison, GRT Race Cars. Race car construction • Mounts/ brackets/interior • Wiring & plumbing • Race car body• Painting • Front suspension • Rear suspensions • Shocks • Braking system • Tires & wheels • Trans & Driveline • Chassis set-up • Track tuning • Repairing a crashed car. **#S234...$24.95**

Paved Track Stock Car Technology
Includes: Step-by-step chassis setup & alignment • Front suspension & steering • Rear susp. systems • Selecting springs & shocks • Chassis tuning with shocks • Scaling the car & adjusting the wt. • Track chassis tuning • & more. **#S239..$24.95**

How To Build & Modify Chevy S/B V8 Cyl Heads
Complete instructions for port modifications & porting techniques, creating air flow efficiency, using a flow bench, high performance valve trains, & more. Covers stock, aftermkt & aluminum heads. **#S252...$19.95**

Pony Stock/Mini Stock Racing Technology
Includes: Complete performance build-up of the Ford 2300 cc engine • Fabricating & prepping a Pinto chassis • Scaling & adjusting the wt. • Detailed chassis setup specs & procedures • Front & rear suspensions • Low buck part sources • Track tuning • For dirt & paved tracks. Covers Pro-4's too. **#S258...$24.95**

I.M.C.A. Modified Racing Technology
An all new book by Steve Smith. Covers: chassis structure & design, step-by-step chassis setup & alignment, chassis tuning with shocks, track tuning, front suspension, all rear suspensions – 3-link, swing arm, 4-link, coil/monoleaf, & more. Everything to get you competitive. **#S280...$24.95** Special Package Price for book & Building The IMCA Video is only $58.95

The Racer's Ultimate Internet Resource Guide
The most comprehensive internet listings of parts mfrs, suppliers, services, and more. **#S276...$10.95**

Paved Track Late Model Chassis Setup VIDEO
With Craig Raudman, Victory Circle Race Cars. Detailed step-by-step chassis setup, with lots of tips & tricks. **#V279...$39.95**

IMPORTANT ORDERING REMINDERS:

- Shipping cost: Add $7 for first item, $8 for 2 items, $10 for 3 or more
- UPS delivery only. Provide street address
- Checks over $35 must clear first (30 days)
- Canadian personal checks not accepted
- Prices subject to change without notice
- In Canada add $5 to all orders
- CA residents add 7.75% sales tax
- US funds only. No COD's
- CALL FOR FREE CATALOG

P O Box 11631-W, Santa Ana, CA 92711

Fax: 714-639-9741 • Web Ordering: www.ssapubl.com • Order Hotline: 714-639-7681